25.95

GV
558
.M34
2014

FASTER, HIGHER, STRONGER

FASTER, HIGHER, STRONGER

How Sports Science Is Creating a
New Generation of Superathletes—and
What We Can Learn from Them

MARK McCLUSKY

HUDSON
STREET
PRESS

HUDSON STREET PRESS
Published by the Penguin Group
Penguin Group (USA) LLC
375 Hudson Street
New York, New York 10014

USA | Canada | UK | Ireland | Australia | New Zealand | India | South Africa | China
penguin.com
A Penguin Random House Company

First published by Hudson Street Press, a member of Penguin Group (USA) LLC, 2014

REGISTERED TRADEMARK—MARCA REGISTRADA
HUDSON
STREET
PRESS

ISBN 978-1-59463-153-5

Printed in the United States of America
10 9 8 7 6 5 4 3 2 1

Set in Adobe Garamond Pro

For Kristen

CONTENTS

INTRODUCTION

Christopher Gore pauses and smiles as I ask him the Question. We're sitting in a conference room at the Australian Institute of Sport in Canberra, where he's the head of the physiology department. We've just spent the past hour digging into the complex process that powers our cells at a muscular level, like our own little graduate molecular biology seminar. I've come more than seventy-five hundred miles to the AIS to try and understand the secrets that the scientists, athletes, and coaches at the institute have uncovered in their decades of work to improve human performance. Getting to pick the brain of someone like Gore, who has done ground-breaking research on altitude training, on drug testing, and on precooling athletes for better performance in hot conditions, is a jump right into the deep end of the pool.

Near the end of our interview, after probing him with questions on all the ways that he tries to optimize elite athletes, I ask Gore what he's learned from all his experience and research that can be applied to everyday athletes. After a moment, he says, "Well, these elite athletes, they aren't like us."

And that's surely true. Part of what fascinates us about the world's great athletes is how different they are from you and me. When we watch LeBron James drive through the lane with his unmatched combination of power and grace, we don't kid ourselves into thinking that we might be able to do the same thing—we marvel at his unique abilities. When we watch Mike Trout crush a fastball into the gap and sprint around the bases, we feel awe at his athletic gifts. When we watch an Olympic skier like Lindsey Vonn barrel down the hill, we shake our heads in respect for her fearlessness and aggression.

Like many writers who are drawn to sports, I'm an athlete myself, although at a far lower level than these stars. When I was a kid, I was an OK bike racer, but never made it to the top ranks. Even when I was riding hundreds of miles a week, I saw other competitors who had invested less time and energy pass me by. Looking back, I realize this may have been what planted the seed for this book in my mind—were they just naturally better than I was? Did they work harder? Or had they found smarter ways to train that offered them greater benefits? Had they found some secret that I wasn't aware of?

In the years I've been writing about sports and technology at *Sports Illustrated* and *WIRED*, I've been chasing the answers to some of these questions. Having covered three Olympic Games and countless other sporting events, I've talked to athletes, coaches, and scientists about what goes into the development of truly great performers, the kind of competitors whom we tell our kids we are lucky to see do their thing.

What I've come to believe is that there aren't any easy answers to the question of what makes a great athlete—so many factors have to align to give us transcendent performers like Serena Williams and Usain Bolt. But there is a thread that unites our best athletes and teams today, and that's an increasing focus on science and technology as a way to push the boundaries of human performance.

Throughout this book, I'll argue that over the past century, we've made massive improvements in the athletic world through a better understanding of our bodies and how they can be trained. But we've seen recently that it's becoming harder and harder to improve at the same rate—it's much more difficult to smash a world record than it used to be. Because of this, athletes have to be smarter about their training—surrounding themselves with a savvy team of scientists and technologists becomes basically essential. Although races can still be won through hard work and effort, they are increasingly won by competitors who not only work hard, but are smarter than the competition as well.

Here's an example from a very different world: Arie de Geus was an executive at Royal Dutch Shell for thirty-eight years, most prominently as the head of the company's strategic planning group. While de Geus was there, Shell became one of the largest companies in the world, partly through his group's innovations in what's called "scenario planning," a technique that helps businesses and organizations develop flexible long-term plans by trying to envision scenarios that might play out in the future. In the mid-1980s, de Geus and his team had research that suggested that the price of oil, then twenty-eight dollars a barrel, might begin to decline, perhaps down to fifteen dollars a barrel (in these days of a hundred dollars per barrel of crude oil, this seems quaint).

De Geus and his team challenged the management at Shell to imagine how they would react to such a situation. When the price of oil plummeted as predicted (and all the way to ten dollars a barrel), the company had the advantage of having considered what to do in a way that much of its competition hadn't. Rather than being frozen by panic, the company was able to quickly change how it evaluated new drilling projects, emphasizing their cost rather than just the expected output of oil. De Geus had a particular saying that encapsulates the lessons he drew from such situations: "The ability to learn faster than your competitors may be the only sustainable competitive advantage."

De Geus's observation was repeated to me by Scott Drawer, the former head of research and innovation at UK Sport, one of the most successful athletic organizations of the past decade. To Drawer, this is the principle on which really great elite-athletic organizations need to be built. You may have better athletes or worse athletes, and that will change over time. But you can still find ways to improve by ensuring that you're always learning as much and as quickly as you can. "It's all about the pace at which you can move. You've got to be willing to take risks and try things out in high performance," says Drawer. "Because even if it doesn't work, you can learn from it. It's the willingness to engage in that process that's most crucial."

De Geus's quote points out that, really, all any of us can do is keep our eyes and ears—and most important, our mind—open to all possibilities in a given situation, with a willingness to try things and learn from them, whether we fail or succeed. All of us, from elite athletes and teams to weekend warriors, can become stuck in our ways and our thinking, and we can find ourselves falling behind competitors who are more nimble, more willing to experiment, and more comfortable with pushing the boundaries.

This is the great joy and torment about working on the cutting edge of science and performance. Today's greatest innovations are tomorrow's baseline, and you have to keep moving forward. That's the only way to continue our physical and intellectual growth as a species; it's the only way we'll continue to run faster, jump higher, and become stronger.

Throughout the years I've spent reporting on this intersection of science and sports, I've found that there are many things that elites do—not just specific tools and techniques they use, but ways of approaching problems—that can be applied to the rest of us. As we explore the world of elite performance, I've tried to highlight those lessons along the way. But as Gore notes, there are also things elites do that only work for them because of their unique physiology or abilities—we'll look at plenty of those things as well.

I hope that as we dive into the world of sports science together, you'll feel the same awe for the scientists and technologists, as well as the athletes, that I do. The myth of the lone athlete toiling away for years in pursuit of a gold medal is a romantic but outdated notion. Today's reality is more complicated and, I think, even more impressive.

FASTER, HIGHER, STRONGER

HACKING THE ATHLETE

Sweat and Science in Silicon Valley

In an anonymous stretch of office buildings off Highway 101 in Menlo Park, California, just miles from tech giants like Facebook and Google, sits a laboratory dedicated to the perfection of athletic training. On this Bay Area January day, the large rolling door of the warehouse-like space is thrown open to let in the sunshine, filling the room with light as a dozen professional baseball players embark on their ninety-minute workout.

They begin by rolling out their muscles to loosen them and increase blood flow before starting the actual work. Rather than using the more typical foam rollers, each player employs a four-foot length of PVC pipe to help free up the tightness in his quads and hamstrings. No traditional static stretching here, no toe-touches. Research shows that doing those activities before exercise can actually decrease an athlete's strength and explosiveness. After the rolling, the players start to move around the gym, gradually ramping up their activity level (this prevents injury, and is a better method for all of us, not just professional athletes, to warm up).

After fifteen minutes, they move on to the next part of their warm-up, although to an observer, it sure looks like they're going all out. The position players take turns working on their sprint technique on an indoor forty-meter track. They take the wide-leg crouch of a runner leading off of first base, and on a coach's prompt, they pivot and sprint toward an imaginary second. As they do, the coach gives them cues on their form, urging them to maintain the correct body lean as they run, so that all of the force they create propels them forward as quickly as possible. Each repetition is timed by an electronic system; the coach enters the time into software on his iPad, where it goes into the player's data file, which tracks everything he does at the gym. A plasma screen TV shows a delayed video feed of the start zone, which players can stop and look at after their sprint to get visual feedback on their performance.

Then the workout proper starts, a blur of activity and action enhanced by the blasting music on the sound system. The players use specific lifts and drills to target their individual needs for improvement. They do one-legged squats to increase their leg strength and core stability, dead lifts to boost their overall strength. Box jumps aim to increase their explosive power, and slide boards are used to work on lateral agility. And because they play baseball, a sport that relies on rotational strength for both pitchers and hitters, they do a special exercise to boost that skill. The player takes a medicine ball, steps onto his back foot and rotates to wind up, then blasts forward and through, heaving the ball sideways against a wall. A radar readout shows them how hard they've thrown it.

Then, somehow, the players manage to amp up the energy even further. The music, a heavy metal song, is cranked as the coaches gather the players into two groups. Everything so far has been mapped and planned to the second; now it's time for a little raw competition. One group heads to the weight racks, where they see who can do the most standing rows in two minutes. The others start doing push-ups with a forty-five-pound weight on their backs, competing head-to-head to see who can complete more. Professional athletes being what they are, there's a little trash talk from some, and a renewed level of focus and concentration from others. The winners try to celebrate, although one is so shattered from his effort that he collapses on a bench and holds his head in his hands, chest heaving, as he tries to catch his breath.

And just like that, they're done. The athletes towel off their faces and head to a kitchen, where they make recovery drinks and slurp them down. The four coaches circulate through the group, offering a look at the data from each athlete's workout and a word of encouragement or correction, depending on what they think the player needs.

This lab is called Sparta Performance Science, and it's the brainchild of a doctor named Phil Wagner. You've probably never heard of him. But you've probably heard of some of his clients. Most notably, he turned Jeremy Lin, a midlevel professional basketball prospect who was having trouble sticking with an NBA team, into a phenom who landed a $25 million contract. Jason Castro, the Houston Astros' all-star catcher? International soccer star Teresa Noyola? The top overall pick in baseball's 2013 draft, pitcher Mark Appel? They've all worked out with Sparta.

Wagner's system is built on testing functional movements of athletes, then trying to optimize them based on individual strengths and weaknesses and how they match up to the sport the athlete plays. Instead of working off of a generalized idea of what an athlete needs to be successful, the lab has identified the specific abilities a player requires to excel in a given sport. A baseball player or golfer needs good rotational ability; a soccer player needs to have great lateral movement; a sprinter needs explosive power in a straight line. Testing reveals where an athlete can improve for his or her sport, and the workout program is developed based on that information.

Sparta isn't alone in this quest—the intellectual spirit behind its approach is something like the spirit that powers innovation in Silicon Valley. It's a systems-based approach, one that tries to boil athletic performance down to its essential parts and then works to improve each one. It's a way of viewing training and sports science more like a software engineer would, rewriting code to improve performance at every turn.

This level of individual attention has been revolutionary in the sporting world, and it points out a fundamental truth of elite athletics in the twenty-first century: Traditional training is no longer enough. To go from being just great to being the best in the world, it's now essential to optimize every bit of performance, even if the gain is just a hundredth of a second. So in addition to relying on their coaches and teammates, athletes work with biomechanists, physiologists, psychologists, nutritionists, strength coaches, recovery experts, and statistical analysts.

Rather than just eating their Wheaties like Bruce Jenner, they guzzle beet juice before a workout, because their team of nutritionists has determined that the nitrates it contains can improve aerobic exercise performance. They don't just rub Bengay on tired muscles; they follow elaborate hydrotherapy regimens to limit muscle damage and reduce soreness. And instead of pounding out hour after hour of training, they do highly targeted workouts like the ones at Sparta, which offer greater benefits for the same amount of effort.

In short, science and technology have become an integral part of an athlete's quest to reach new frontiers of accomplishment. Assembling a group of sports scientists and coaches used to be beyond the realm of possibility for most athletes, but now the existence of national sports institutes and commercial sponsors allows it to happen regularly, across any sport you can think of.

From the Olympics to the NBA Finals, from the World Series to the Tour de France, from high-tech labs in Canberra and Colorado Springs to converted warehouses in Santa Monica, California, this revolution has changed our games forever. It took evolution millions of years to produce the modern human. Over the past hundred years, coaches have used intuition and discipline to vastly improve athletic performance. Now scientists are taking the next step, helping athletes approach perfection. This book is about that journey.

The Medals Are in the Details

By definition, elite athletes are, well, elite. They're rare. They're different from most of us—physiologically, psychologically, genetically. There's something about them that has allowed them to reach the very pinnacle of human performance, some mixture of talent and effort, gifts both bestowed and earned that allow them to do things about which most of us can only dream.

"In the United Kingdom, we invest in about fourteen hundred athletes out of a population of sixty-five million," says Scott Drawer, who headed up UK Sport's research and innovation arm in the lead-up to the 2012 Olympics in London. "You're talking about the true extremes of individuals, a population that we understand little about."

But the key to winning Olympic gold is to somehow become the elite of all the elite, standing at the pinnacle of a very tall pyramid. Where do you find the edge that will put an athlete at the top of the podium listening to her national anthem with a gold medal around her neck? It turns out you do it by focusing on the little things. Just look at the example of British cycling.

For most of its history, the British cycling team wasn't just a laggard; it was completely noncompetitive. Between 1924 and 1988, British cyclists won exactly zero gold medals at the Olympics; from 1960 to 1988, they won only two bronze medals. Chris Boardman won a cycling gold for Great Britain at the 1992 Barcelona games, but the country was still nowhere on the world stage, where the sport was dominated by countries like France and Germany.

But by 2012, the British had become the most successful cycling nation on earth. Just before the Olympics, Bradley Wiggins had become the first British rider to win the Tour de France (riding for British-run Team Sky), and he went on to win gold in London in the individual time trial. The next year, Wiggins wouldn't defend his Tour title, but that was no bother. Team Sky was instead led by a rider named Chris Froome, who also won the race. At the London Olympics, British riders would win twelve total medals in road and track cycling, twice as many as any other nation. The former laughingstock had become seemingly unstoppable.

The story of British cycling isn't one of luck or even plucky can-do spirit, but rather one of smart athletes, coaches, researchers, and scientists all working together to hack the hypercompetitive world of elite sports.

Through hard work and clever innovations, breakthrough technology and dogged attention to detail, they came to dominate a sport.

The guiding principle behind the British juggernaut is a simple phrase coined by Dave Brailsford, who until April 2014 was the performance director of British Cycling, the sport's governing body in the country, as well as the general manager of Team Sky. Having overseen the dominance of British cycling for the past decade, he's managed to boil down the philosophy behind all that success—the gold medals and world championships and yellow jerseys—into a single phrase, which he shared with Richard Moore in his book *Heroes, Villains & Velodromes*: "Performance by the aggregation of marginal gains."

What does Brailsford mean by the "aggregation of marginal gains"? It's a reaction to where we are in the development of sports. We spent the twentieth century learning the basic science of human physiology, training, and nutrition. In that time period, sport went from a largely amateur pastime to a worldwide industry worth hundreds of billions of dollars, and millions of new athletes were able to compete. Technologically advanced new equipment—from stronger, lighter shoes to drag-reducing swimsuits—radically elevated performance. Our understanding of nutrition led to the development of drinks and foods that ensure athletes have the fuel they need during events. Strength and conditioning programs meant fewer injuries and more time to train. In a hundred years, we developed a cohort of athletes whose physical accomplishments would have been unthinkable four generations earlier.

The challenges of the twenty-first century are very different. At this point, we've made most of the big leaps in our natural physical capacity. That's what Brailsford is acknowledging when he talks about marginal gains. Instead of looking for one earth-shattering change, British cycling takes a different approach. It looks at every aspect of performance, and tries to improve each a little bit—even just a tenth of a percent. If you find a training technique that makes an athlete that tiny bit stronger, it alone might not have a huge effect on a race. But if you can stack those very small improvements on one another, finding a bit in tires and a bit in the wheels and a bit on the track surface and a bit in nutrition supplements—well, soon those marginal gains begin to add up to big gaps between you and your competition.

Some of these differences can be found between individual athletes—no one wins an Olympic gold medal without exceptional physiology and without having worked demonically to maximize their potential. But those are some of the things we've worked the most to understand. Sports scientist Giuseppe Lippi and his coauthors have written that in the future "athletic performance will be determined less and less by the innate physiology of the athlete, and more and more by scientific and technological advances." Obviously, you still need to find and train the best athletes, but you need to have the best PhD's on board as well.

The marginal gains philosophy requires you to look at every single aspect of what you do so you can try and improve it. So, for instance, the British cyclists all travel to competitions with their own pillows, to ensure that they're as comfortable as possible and eliminate any potential neck problems. There's an obsessive but understandable focus on hand washing and hygiene: An International Olympic Committee study found that 7 percent of competitors at the London Olympics suffered from some sort of illness. All it takes is a cold on the wrong day to obliterate a lifetime's hard work. And before races, the team sprays its tires with alcohol, making them slightly tackier and increasing their grip at the standing start.

These marginal gains are so important that British Cycling has actually created a position called "head of marginal gains," which was held during the London Olympics by sports scientist Matt Parker. He coordinated a staff of fifteen, all of whom were looking for those little edges. One good example from London was what the team called Project Golden Hour, which came from a close examination of the schedule for the women's team pursuit event. A year before the competition, when the schedule was released, Parker and his team noticed that there was a very short window of just an hour between the semifinals and finals of the event, so they set out to develop ways to maximize rider recovery in that hour. "And if you look at all the teams' data," Parker told the *Independent*, "we're the only ones that went faster in the final of the women's team pursuit than in the semi-finals," a ride that resulted in a gold medal and a new world record.

The Fundamentals

One of the things that you notice when you start to hang around elite athletes, coaches, and sports scientists is that they think about being an athlete very differently than the general public does. When you watch a sporting event on TV, you see the whole of the activity—the quarterback dropping back, scrambling to avoid a sack, and throwing a pass downfield. There's a tremendous amount of athleticism involved.

But you can break those feats down to a much more fundamental level. Instead of thinking like a fan, scientists think about these things like scientists—physicists, more precisely. Because, as any physicist will tell you, *everything* can be viewed through the lens of physics, including sports. Once you approach sports and athletic movement this way, you can see that there are three building blocks on which every athlete stands: strength, speed, and endurance.

When we think of strength in an athletic context, most of us think of sports like football or weightlifting, in which much of the training is done with weight or resistance work. Of course, strength is a huge asset in these sports. But it's not just about being able to lift weights.

Let's roll it back to a more fundamental level. Strength is a person's ability to apply force. If you're deadlifting three hundred pounds in the gym, you're applying force to the bar to overcome gravity and lift the weight. But athletes apply force in myriad other ways. When a basketball player jumps, she's applying force to the ground to propel herself into the air. When a cyclist rides a bike, he's applying force to the pedals to turn the crankset and move the bike forward. When a baseball player takes a swing, he's applying force to the bat, and then applying that force to the ball (unless, of course, it's a swing and a miss).

Even athletes whom we don't think of as being particularly strength oriented, like runners, actually generate tremendous amounts of force during a race. Research has shown that the difference between running speeds isn't due to the speed with which a runner can move his legs—what is called the recovery time of a stride. In fact, just about all sprinters, from Usain Bolt on down to you and me, swing their legs at roughly the same speed, taking about a third of a second to get from one stride to the next. Instead, the key factor is how much force a runner hits the ground with. In a paper published in 2000, Peter Weyand and his team used a treadmill equipped with devices that detected the force with which runners struck the ground as they ran. As the speed of the treadmill increased, there wasn't an increase in how fast the runners' legs moved as they ran. What changed was the force with which the runners were hitting the ground: "Speed is conferred predominantly by an enhanced ability to generate and transmit muscular force to the ground." Running faster means applying more force. (This is a good argument for runners to spend more time doing weight training, which many recreational runners skip altogether. A 2014 meta-analysis found significant improvements in performance for distance runners and other endurance athletes after strength training.)

We're talking about a lot of force. Sprinting full out, Usain Bolt applies up to roughly one thousand pounds of force to the track when his leg makes contact with the running surface—far more than he could manage doing a one-leg squat at the gym. (An average man sprinting all out would likely generate only about half the force that Bolt does.) Ralph Mann, a biomechanist who has worked with many top U.S. sprinters, said at a 2012 coach's clinic that to have a good burst off the starting blocks, a runner needs to be able to do squats in the weight room at six hundred pounds or more.

Strength, when viewed in this context, is the foundational aspect of athletic performance. Without applying force, there's no movement at all. All other things being equal, in just about every sport you can think of (outside of perhaps shooting and archery), the athlete who can apply more force than his competition in the right situation will win.

If strength is a measurement of how much force can be generated, speed is a measurement of how quickly a movement can be completed. There are two main factors that feed into the ability to complete a movement more quickly. The first is the composition of the muscles involved in the movement, as some types of muscles are capable of contracting more quickly than others. The other is the neuromuscular component to speed, which is very important.

We humans have two major types of muscle fibers in our bodies: fast-twitch and slow-twitch. Without diving too deeply into the differences, fast-twitch muscle fibers are able to contract faster than slow-twitch fibers. That means that an athlete with a larger proportion of fast-twitch muscle fibers will be able to complete a movement over a shorter amount of time than someone with a larger proportion of slow-twitch muscle.

But there's also a neuromuscular component to speed. Over time, you can train your nervous system and muscles to complete movements more quickly. Think of a pianist who's learning a new song. At first, the movements of the fingers and arms necessary to play the song aren't familiar, and the performance is halting and awkward. But with practice and repetition, she starts to build connections in the neurons and the brain that allow her to play the piece faster and faster.

The same thing happens for athletes. A beginning hockey player's wrist shot might take seemingly forever to complete, but an NHL player can have the puck flying at the goal in the blink of an eye. The move hasn't become more complicated; he can simply execute it faster.

Of course, speed and strength interact with one another; when they do, that's what we call power. Power is an expression of how quickly one can apply a specific force. Or to express it in terms of the other two:

Strength x Speed = Power

Take the two sports of Olympic weightlifting and powerlifting. In the two Olympic lifts—the snatch and the clean and jerk—the bar is basically thrown (carefully!) into the air by the athlete, who then positions himself under it to complete the lift. Power lifts such as the squat and dead lift don't require the same sort of explosive movement. Obviously, an Olympic weightlifter has a very high level of strength, but powerlifters are often stronger. The difference occurs in the arena of power. Olympic lifters are immensely powerful—they can apply their (relatively) lower level of strength much more quickly than powerlifters can. (It might be more accurate to call powerlifters "strengthlifters" instead.)

That preceding equation shows an essential truth of many athletic movements. There are two ways to increase how far you hit a golf ball or throw a baseball. You can increase your strength, which will help you apply more force during a movement, or you can increase your speed, so that the same force is applied more quickly. And if you want to make it to the very top of the athletic world, you optimize both. Maybe you're a golfer who wants to hit your driver farther off the tee. Start in the gym, where you can concentrate on exercises like the medicine ball throws at Sparta that build rotational strength. And then hit the driving range, working to improve your neuromuscular speed so you can take best advantage of that new strength.

OK, so we have considered how much force you can apply, and how quickly you can apply it. Now let's think about how long you can apply the same force again and again without slowing down or losing steam—or, to put it another way, endurance.

Endurance is massively complicated. Everything from our cardiovascular fitness to nutrition to metabolism to size has a huge effect on it. We'll look at many of those factors on their own later in the book, but they all feed into the same core ability to perform a movement at a sustained rate with relatively little degradation. Many reasonably fit people can climb off the couch and sprint a hundred meters in seventeen seconds or so. That same pace would allow you to break the two-hour barrier in a marathon—if you could do it another 422 times in a row. That's endurance.

There are other factors that play a huge role in sporting success, from tactical and strategic awareness to psychology to nutrition, and we'll look at all of them over the course of this book. But when we talk about athleticism, we're really talking about these three pillars—strength, speed, and endurance—and the balance between them for any particular sport.

The trick is, some of these factors operate in opposition to one another. Very broadly, strength and endurance aren't aligned. Think of a marathoner and a weightlifter—this is how far apart they are on the spectrum. The hugely strong lifter doesn't have a lot of endurance, and the unstoppable endurance runner doesn't have the ability to hoist five hundred pounds over his head.

So the science (and art) of training and coaching is to try and get the balance of all of these things just right. Spending time and energy to develop unnecessary capabilities in an elite athlete can be really problematic—that's time you're not spending on things that *are* essential. But beyond training, athletes can be hacked by other means, ones that touch on their bodies and minds in ways that we're still trying to fully understand.

Load and Explode

Back at Sparta Performance Science, I'm getting ready for my performance testing—the same test that the company does with clients like Major League Baseball's Colorado Rockies, and the National Football League's Atlanta Falcons and Jacksonville Jaguars. The test itself is simple: I'll step onto a contraption called a "force plate," which measures the impulses that come through my feet onto the ground. I'll do six vertical jumps as high and explosively as I can, and then Sparta's software will crunch all those numbers and give me what they call a "movement signature."

That signature will break down three different aspects of the force I apply when I jump. The "load" captures how quickly I can start to develop force. "Explode" is a measurement of how well I can transfer that force. And "drive" tells us how long I can produce that force, how I finish the movement. Think of the three of them as a chain that links together as I jump.

I do my six trials, and then go over to the computer, where Phil Wagner waits for the results to be calculated. It turns out that my movement signature shows acceptable load and explode, and very high drive compared with the other two variables. At Sparta, they call this sort of result (in which one variable is way out of whack from the other two) an "extreme" signature; in their lingo, I have an extreme drive movement pattern.

Wagner runs down what they know about athletes with this movement signature. "These sorts of athletes tend to excel in sports or situations that revolve around timing," says Wagner. "You'll see basketball players who are good on the pick-and-roll with this." A high drive means that you're able to produce force over a longer period. Consequently, you're able to time its application more precisely. "You're also in increased danger of knee and lower back pain," Wagner says.

"So, what would you do with me if I was one of your athletes?" I ask Wagner.

"Well, that depends on what sport you want to play," he replies. "If you were looking to run faster like a wide receiver, we'd look to increase your explode."

I tell him that I've been trying to improve my golf game. Wagner suggests I work on my load—focusing on the front of my body like the ankle, knee, and quad, and working to increase their strength. "That's where force generation begins," he says.

The testing is just part of what Sparta does. The other part is turning those test results into training plans, based on the movement patterns of the athlete and the sport in which he or she is looking to compete. I could spend lots of time trying to increase my explode, for instance, but it's not really an athletic skill I need (wide receiver not being one of my current goals). There's only so much time and energy available to all of us, elite athlete and weekend warrior alike. Anything you decide to do takes away from your ability to do something else—an hour watching TV means an hour not spent at the gym. Wagner and his team are trying to build a system that ensures that all of their athletes are focused on the biggest bang for the buck in every single aspect of performance.

There are myriad factors that enter into athletic success. Some of them, like Wagner's insight into time and energy, are applicable to all of us. Some of them are only really relevant at the uppermost levels of competition. After all, the tiny percentage of improvement I might see in my weekend bike-riding speed by spraying the tires with alcohol pales in comparison to what I could see if I just, you know, trained harder and got myself in better shape.

In a scientific paper, N is shorthand for the number of participants, or the sample size. There's a lively debate in science about how statistics are used, how big a sample size needs to be, and how significant the variations between groups need to be to truly demonstrate an effect in a study. But generally, more subjects—a bigger N—is better. The larger the group of people you try something with, the more authoritative you can be with your conclusions.

All of us who are trying to improve athletically—from gold medalists on down to you and me—are involved in an ongoing experiment with one subject. Or to put it another way, N=1. That's why Wagner's work to match up a specific athlete's movement patterns with a specific sport's requirements is so powerful.

Sport is an experiment that we're all participating in, from fans to scientists to coaches to the athletes themselves. We haven't been doing it for that long, evolutionarily speaking, but we've made massive progress thus far. That progress seems to be slowing, but part of the reason for that is that we've been treating the experiment like a group project, when in fact it's an extremely personal one. And the reason for that individual variation can be found inside each of our bodies—in our DNA.

2

GOLD MEDAL GENETICS

DNA Isn't Destiny, but It Sure Can Help

Growing up in Quebec City, Canada, Claude Bouchard loved ice hockey, which is no surprise. There might not be another pairing of a sport and a nation that's as deeply entwined as hockey and Canada. The country defines itself through the game, and vice versa. Canadians think the traits that help you win hockey games—toughness, hard work, humility, determination, strength—reflect the best of their country. And the world of hockey, even at the highest level, takes on the humble, self-deprecating, slightly stoic spirit of Canadians. It's a unifying force in a country that's often found itself separated along geographic or linguistic lines. It's not that Canada cares about hockey. It thinks it *is* hockey.

So like so many Canadian boys before him and since, Bouchard started playing the game at a young age, working his way up through the levels of the sport, from Tyke to Novice to Atom to Peewee to Bantam. He loved the sport, and he was good at it. But then something happened, or more correctly, didn't happen.

"Guys were getting bigger and bigger," says Bouchard in his French-inflected accent. "They became bigger than I was—my rate of growth slowed down, but not theirs. It was OK for a couple of years, but when I moved to Junior hockey at seventeen or eighteen years old, the guys were so big that I would have gotten killed. I became realistic."

As that young hockey player, Bouchard had run up against a stark truth, one that was heavily influenced by genetics. The general consensus on height is that it's about 70 to 80 percent heritable—that means that about three-quarters of the differences in people's heights are attributable to genetic differences between them, while the remainder is due to environmental factors, most notably the quality of their nutrition.

Bouchard didn't know this then, at least in these terms. He might have noticed that kids who kept getting taller tended to have taller parents, but he wouldn't have been able to explain the mechanics of heritability, nor would he have been able to explain the complexity of that genetic role. There's no single gene that controls height; scientists have identified thousands of variations that seem to influence it in a tangled, complicated web of relationships and effects. What Bouchard did know was that when he stopped getting taller, his hockey career was over. He switched to other recreational athletic pursuits, like skiing.

He headed to college at Laval University, in his hometown, and started to study medicine and genetics. He worked in cardiology for a while, but there was an idea that kept tugging at him from the back of his mind. "I've always been interested in human performance," he says. "I think it was that involvement I had in sports and how much I enjoyed it. It stayed with me." Over time, Bouchard decided to organize his academic career around that interest.

He began by getting funding for some small studies around the heritability of physical traits, using families and twins. Throughout the 1980s, he used these studies to suggest that there was a strong possibility that many traits beyond just the physical, including our responses to exercise, had a strong genetic component. But Bouchard had run up against the limits of his sample size and the amount of data he could collect. He needed a much larger group to study, and he needed to gather much more information about them. By this time, it was becoming possible for scientists to study DNA more closely, and Bouchard envisioned a study in which all the genetic material for each subject would be collected. He put together a consortium of five institutions, got funding from the National Institutes of Health, and launched what he later named the HERITAGE Study (HEalth, RIsk factors, exercise Training, And GEnetics).

To get funding, Bouchard and his team included a lot of information about various diseases and health outcomes—he says that it's very hard to get financing for research on performance without that focus. But if you read the published paper that outlines the study (tellingly, in an American College of Sports Medicine journal), it's clear what they really wanted to explore:

> The overall objective of the HERITAGE project is to study the role of the genotype in cardiovascular, metabolic, and hormonal responses to aerobic exercise training and the contribution of regular exercise to changes in several cardiovascular disease and diabetes risk factors.

With HERITAGE, Bouchard hoped he would finally be able to answer a question that had nagged at him for years: How much of our athletic ability comes from our DNA?

All in the Family

The idea that our athletic abilities are genetically determined to some extent is both completely apparent and highly disputed. In some settings, we totally accept these genetic differences without a second thought; in others, we find the suggestion of their existence not just controversial, but offensive.

Take, for instance, the prevalence of multigenerational success at the highest level of sports. Making one's way into the NFL, for instance, is a very, very hard thing to do, but an awful lot of families have multiple members who have played pro football. The Pro Football Hall of Fame tracks fathers and sons who have played pro football—as of 2014, there have been 206 sets of fathers and sons who've played in the NFL and other pro leagues, with a total of 221 second-generation players. There have even been five third-generation players, led by the seemingly endless Matthews dynasty (currently headed by Green Bay Packers star Clay Matthews III and his brother, Philadelphia Eagles linebacker Casey Matthews).

When it comes to siblings, 364 sets of brothers have been pro football players, a very high proportion of the overall population of NFL players. As Australian sports scientists Kevin Norton and Tim Olds wrote when analyzing family relationships in the NFL, "The probability of this pattern of familial selection happening by chance is infinitely small." There are certainly other factors at play here—the son of a professional football player is likely raised in an environment where he can encounter the game at a young age and perhaps gain insights from his father. But given the physical demands of the game, it's not shocking to see those familial patterns. Football is a game played by large men, and large men are more likely to have large offspring.

There's also the matter of what Norton and Olds call "assortative mating," in which athletes marry one another. They point out that this athletic coupling "often leads to the production of genetic polymorphisms of the next generation of gifted athletes." A less academic analysis of this pattern comes from Ashton Eaton, the world record holder in the decathlon. Eaton met his wife, Brianne Theisen, when they were both competing for the University of Oregon; she's a world championship silver medalist in the heptathlon. When asked why athletes get better and better in the track and field world, Eaton says, "We keep marrying each other." It's a joke that contains a grain of truth—if I had to bet on a kid to do pretty well in a track meet, I could do worse than their offspring.

There is one genetic factor whose influence on sports performance is so profound that we organize separate competitions to account for the differences between athletes. I'm talking, of course, about sex. It's a division that's so ingrained in most sports to the point that we don't even think of it—at the Olympic level, men and women only compete directly in two sports: equestrian and sailing. But it's useful to walk through sex as an example of the influences of the genome on athletic performance.

Men and women differ in several physical and physiological ways that have a direct impact on athletic performance. The average woman is 5 inches shorter and 30 to 40 pounds lighter than the average man. She also tends to have about 6 to 10 percent more body fat. She's not as strong, either—about 40 to 60 percent weaker in the upper body and 25 to 30 percent weaker in the lower body. She's likely not as fit at a cardiovascular level, with a VO_2 max (a measurement of an athlete's aerobic endurance capacity) that's about 85 percent of the average man's. All in all, there's a sizable physiological gap between men and women. (We're talking about men and women at similar levels of training here—trained women obviously are much more fit than poorly trained men.)

How big a gap? French researchers built a database of the world records and ten best performances by men and women on an annual basis across eighty-two events in five sports, including track, swimming, speed skating, cycling, and weightlifting. After crunching the numbers, the team found that the average difference between world-record performances of men and women across all these events is about 10 percent. In some specific events, the gap is smaller, as low as 5.52 percent in the 800 meter freestyle swimming event. The gap of the ten best performances in these events is a little larger, about 11.7 percent. And while some researchers have hypothesized, given the more rapid improvement of women's records in the recent past, that someday women may outrun and outswim men, the likelihood of that outcome doesn't seem to be borne out in the actual numbers—the sex gap between men and women has been fairly steady since 1983. "Without any technological improvement specifically dedicated to one gender or the other," the researchers write, "performances will probably evolve in a similar manner for both men and women."

This is all to point out that there's a massive and seemingly stable difference in performance based on sex—and sex is completely genetically determined. If you have the male XY chromosomes, before you're even born you've been given a massive athletic advantage over a sister, who has the same parents and a set of XX chromosomes. And given the huge systemic differences that a simple chromosome switch has on traits that affect athletic performances, shouldn't we expect to see those differences around other genetic variations?

Where this gets even more complicated is in cases like that of South African runner Caster Semenya. Semenya won the 2009 World Championship in the 800 meters, but drew attention for her seemingly male features. She was forced to undergo gender verification testing by international track officials, kicking off a yearlong process of sensational news stories and rampant speculation. She was later cleared to compete as a woman. While the results of her testing haven't been made public, there have been media reports that she has both female and male characteristics—that she's intersex, a condition in which chromosomal sex and primary sex characteristics, like genitals, differ. Purportedly, that gave her a higher level of testosterone than her competitors. The speculation is that she's been given treatments to bring those levels in line with other women's. This is obviously an extreme case, but also a good reminder that sex isn't quite as binary as we often imagine, and can have a big impact in an athletic setting. (Semenya went on to win a silver medal at the London Olympics, competing as a woman.)

Do These Genes Make Me Fast?

If separating male and female athletes based on their genetic ability is a completely common practice, imagine a different possibility much likelier to ruffle feathers. What if we were to separate certain athletic events based on race? After all, in some events, there seems to be a clear pattern of superiority at the upper levels of competition by some specific racial and ethnic groups. Perhaps the most prominent of these examples is the nearly complete dominance of distance running by athletes from East Africa, specifically Kenya and Ethiopia.

The statistics are familiar to many sports fans, but still impressive. As of March 2014, fifty-one men in history have run a sub-2:06 marathon. Of those runners, forty-seven are Kenyan or Ethiopian. In total, Kenyan and Ethiopian runners account for 90 percent of all the world records and top ten performances in middle- and long-distance track events. It looks for all the world like these runners have some sort of advantage that has vaulted them past their competition—why not allow them to run for one set of medals at the Olympics, and then have separate races for runners who don't hail from those countries?

The recent success of Kenyans and Ethiopians isn't the first illustration of distance running being dominated by runners from a particular area. For decades, Finland was the epicenter of distance running. Finnish runner Paavo Nurmi won 9 Olympic gold medals in the 1920s; in the five Olympics from 1920 to 1936, Finns won 10 of the 15 available Olympic medals in the 10,000 meters and 9 of the 15 in the 5,000 meters, a stunning performance. Was the remarkable success of Finnish runners somehow due to a genetic advantage that they shared because they were Finns?

Well, no. Or not at least one that was ever identified. One of the key factors in their success was the Finns' recognition of the power of interval training, which became standard practice around the world in the decades that followed. But if you were a competitor who didn't know about that training technique, the steady stream of Olympic medalists from Finland might have looked like the result of Finnish stock.

The point is, we aren't very far along in our understanding of why Kenyan and Ethiopian athletes have been so dominant in distance running over the past two decades. But jumping to the conclusion that it's purely genetic is premature, and falls into the trap of writing off all differences in athletic performance to genetics. We seem to be constantly searching for a clear, easy answer to why some athletes perform better than others, and as we do so, we lose nuance and understanding. The dangers of the oversimplified notion of what actually leads to great athletic performance could be why Bouchard downplays some of the claims that have been made around so-called "sports genes" like ACE and ACTN3.

The ACE gene codes for something called angiotensin converting enzyme (hence its name), which helps control blood pressure in our bodies. More ACE in your blood will constrict your blood vessels, constraining circulation and increasing blood pressure. There are two common variations in the ACE gene—what's called an insertion/deletion polymorphism. People with two copies of the insertion variant have less of the enzyme in their blood, while people with two copies of the deletion variant have more.

ACE was the first gene identified as having an impact on athletic performance, and therefore has been extensively studied. To date, the results of those studies seem to indicate, on balance, that the insertion variant can lead to increased performance in endurance sports reliant on moving as much blood as possible through the circulatory system (like marathon running). The deletion variant is associated with increased performance in sprinting or power sports. (There are some studies that don't support these conclusions, but the general consensus is that ACE variation does have some effect.)

But there's one important caveat here: Those observed benefits of the ACE insertion have all been in Caucasian athletes. When researchers have looked at elite endurance runners in Kenya and Ethiopia and compared them with national class runners or control populations, they haven't found any differences in the ACE polymorphism.

That same pattern of difference between ethnic groups shows up in the other most-studied gene in sports performance: ACTN3. That gene encodes for a protein (alpha-actinin-3) that's used in fast-twitch muscle fibers, the sort of muscle that sprinters rely on for their extreme power. There are two common variants of this gene: the normal version, called 577R, which allows you to form the protein in your muscles; and a version called 577X, which prevents you from making the protein. If you have two copies of the 577X version, one each from your mother and father, you won't make any alpha-actinin-3 at all.

Not having the protein doesn't seem to affect people in their daily lives. But it does appear to have an effect on the potentially explosive power your muscles can generate. The results of the first study on this, in 2003, have shown that the 577R variant is associated with high-level sprint performance, compared to someone with two copies of the 577X variant. The researchers hypothesized that increased levels of the protein allowed for more forceful contractions of the muscles, and therefore increased the sprinter's speed.

Interestingly, different populations have different levels of variation in the gene. In the original Australian study, 18 percent of the control population had the XX variant, which means they produced no alpha-actinin-3. That's a big chunk of Australians who would be at a disadvantage for sprinting. Contrast that with a study of Jamaican sprinters, in which researchers found that the control population in the country only contained 2 percent of people with the XX variant. And 75 percent of both the control population and the elite sprinters tested had the RR variant, which has been theorized to be most beneficial for sprinting.

This is some of the inherent complexity that we must consider when attempting to make sweeping generalizations about the relative genetic ability of ethnic or racial groups. In Australia, having the RR version of ACTN3 might separate you from your peers in terms of sprint ability; in Jamaica, it probably wouldn't make a difference, because of its high prevalence in the general population. Based on the genetic evidence, you can't say that Jamaicans are better sprinters than Australians. But you might be able to say that *a larger proportion* of Jamaicans than Australians are likely to be disposed to be great sprinters. The genetics that allow for world-class performance in sprinting probably exist in both countries, but based on what we know today, more Jamaicans probably fit that profile.

While genetics is undoubtedly crucial for an athlete to reach the very top of the sporting world, it isn't the entire story by any stretch of the imagination. The environmental factors involved—from an interest in sprinting to access to coaching to a culture that's heavily invested in sprinting success—play a huge role, as does an athlete's work ethic and training program.

The same holds for East African distance running. We'll talk about some of the factors in play later, but there's a culture of running, a tradition of running heroes, and the fact that the tribes that have produced most of these champions live at high altitude. Being an Olympic champion is such an unlikely thing to be—like snowflakes, each athlete is a unique combination of genetic ability, practice and training, and circumstance. So while we haven't found clear genetic markers that explain East African running dominance (or any other athletic success), we know that a high proportion of athletes there have the body type most advantageous for distance running. That physicality—their phenotype as opposed to their genotype—is crucial. The phenotype best biomechanically suited to distance running (long thin limbs, low fat) is not limited to East Africa, of course, but it certainly is a lot more common in Kenya than Kentucky.

Maxing Out

So just how do a person's genes regulate his or her ability to become a great athlete? Let's do a case study in the world of endurance sports: cycling, distance running, and cross-country skiing. At the 2012 ACSM conference in San Francisco, Claude Bouchard walked the audience through the factors, including genetic ones, that would allow a particular individual to excel in one of these sports.

First, Bouchard noted that his lab had collected data on 310 elite endurance athletes, and specifically that they had measured the VO_2 max of those athletes. VO_2 max is a measurement of how much oxygen an athlete can process during exercise—a higher number is better, indicating that the athlete can take in more oxygen and get it to her muscles. If an endurance athlete is a car, her VO_2 max is the horsepower of the engine.

In Bouchard's previous studies, the average untrained male right off the couch had a VO_2 max of 41 ml/kg/sec—he could use 41 milliliters of oxygen per kilogram of body weight per second. That's the standard for an untrained man. In comparison, the 310 elite endurance athletes had an average VO_2 max of 79 ml/kg/sec, meaning that they could process nearly twice as much oxygen as the untrained men could. Of that elite cohort, 68 percent had a VO_2 max between 75 and 80, 25 percent between 80 and 85, and 7 percent at about 85.

(There's some controversy about which athlete has the highest tested VO_2 max ever. The most frequently given answer to that question is Norwegian cross-country skier Bjørn Dæhlie, who reportedly had a VO_2 max of 96. With that aerobic engine, Dæhlie won twenty-nine medals at the

Olympics and World Championships, including eight Olympic gold medals. Greg LeMond, the American cyclist who won three Tour de France titles, reportedly had a VO_2 max in the nineties as well.)

Now, there's not a perfect correlation between an athlete's VO_2 max and his ability to ski or run or swim or ride fast. There are plenty of elite endurance athletes who win tons of races without the otherworldly VO_2 max numbers that someone like Dæhlie had. In fact, once you get to the elite level, VO_2 max isn't a particularly great predictor of performance. At the very highest levels, things like your anaerobic threshold and economy of movement serve as larger differentiators. If you have a higher threshold, you're able to use a larger percentage of your total capacity for longer—you can rev whatever size engine you have higher. And if you have good economy, you're able to use relatively less energy to go the same speed as a competitor—a little like getting a more fuel-efficient car.

For an elite endurance athlete, a high VO_2 max is more like the price of admission. Without a certain level of sheer oxygen-processing capability, you just won't be able to compete. It might not be the main determinant of who wins at the Olympics, but it's unlikely you'll get into the club without meeting the threshold.

There is not a lot of published data on truly world-class endurance athletes that lets us understand how these factors interact. But there's one remarkable case study available that demonstrates their interaction at the absolute pinnacle of distance running. Written by Andy Jones at the University of Exeter, in the UK, it outlines the physiology of Paula Radcliffe, the women's marathon world record holder. Jones began working as Radcliffe's physiologist in 1991, when Radcliffe was a seventeen-year-old runner in Bedfordshire, England.

"Paula had started to run poorly for some reason," says Jones. "Back in those days, there were no commercial labs, there was no English Institute of Sport where athletes could go and do testing." Jones was a very good runner himself—in fact, he formerly held the British junior records for 10K and the half marathon. He was still running competitively, but had started his studies for a PhD in exercise physiology at the University of

Brighton. Jones's coach knew Radcliffe's coach, and suggested that she visit Jones for testing.

"We discovered why she wasn't running particularly well," says Jones. "Her hemoglobin was quite low; she was somewhat iron deficient. After we identified that and corrected it, she bounced back and won the World Junior Cross Country Championships." After that success, Radcliffe and Jones embarked on a relationship that lasted the rest of Radcliffe's career. Over all those years Jones tested her regularly, ultimately giving us a look at the development of one of the best distance runners of all time.

We've already talked about Bouchard's research showing that elite male runners tend to have a VO_2 max greater than 75. Female athletes have lower VO_2 max levels in general; the range for elite female distance runners tends to be in the 60 to 75 range. From 1992, when Radcliffe was an eighteen-year-old, to 2003, when she set her marathon world record of 2:15:25, her VO_2 max ranged between 65 and 75, usually hovering around 70. As Jones writes, "It should be noted that a VO_2 max of this order is extremely high, even in elite female athletes, supporting the view that a high VO_2 max is a prerequisite for successful performance at the international level." When she was eighteen, and running twenty-five to thirty miles a week—an amount that many dedicated recreational runners put in—Radcliffe had a VO_2 max of approximately 70. There's no getting around it—she was genetically endowed with a very, very high ability to process oxygen.

To have world-class physiology, Bouchard reasons, you need two things: You need to start with a high baseline VO_2 max before you start training, and you have to respond very well to training so you can increase it even further. As he says, "You can't get to an eighty if you start off with a thirty-five."

Remember that in Bouchard's study, the average sedentary male had a VO_2 max of 41. But between 1 in 10 and 1 in 20 of the study group began with a much higher value—above 50. That's a significant difference. And while part of that variation is random chance, Bouchard has found that about 50 percent of it is heritable. Half of our native ability to process oxygen, our foundational endurance capability, comes from the 500,000 or so genetic variations in each of us that are unique to our familial lineage.

For Jones, Radcliffe's test results were actually life changing. "You could see straightaway how talented she was and what she could achieve," he says. "At that point, I still had high ambitions as a runner. I had wanted to be the Olympic champion. But working with Paula made me realize straightaway that I wasn't going to do that. She kind of finished my athletic career off." It was a case of Jones seeing what truly extraordinary physiology looked like, and knowing that he'd never be able to reach that level.

So, let's go back to our hypothetical athlete with a high baseline VO_2 max. How do we get him from 50 to 75 or 80? A 50 percent increase in VO_2 max through training is a tall order. In Bouchard's studies, the average participant saw a 16 percent increase after a standardized twenty-week fitness program. But that average conceals a huge variation in individual response to the program. Some people's VO_2 max improved by less than 5 percent, while some of the other subjects' increased by 30 percent. Again, there appears to be a significant genetic factor here—our trainability for VO_2 max appears to be about 47 percent genetically determined.

Interestingly, Bouchard didn't find any correlation between a high baseline and high trainability. They're two independent traits that have different genetic mechanisms. So you can have four different classes of potential endurance athletes:

High Baseline Low Trainability	High Baseline High Trainability
Low Baseline Low Trainability	Low Baseline High Trainability

You can have a low baseline and low trainability—call it the worst of both worlds. Or you can try to balance high trainability with a low baseline, or vice versa. Maybe that's the case with most of us—we can turn ourselves into OK athletes, but nothing approaching elite. But then there's that magical upper right-hand corner, where you have both the high baseline and the high response to training. What are the odds of winning that particular genetic lottery?

Bouchard estimates that between 1 in 30 and 1 in 50 people have the trainability to allow for a 50 percent increase in VO_2 max above their sedentary level. Combine that high trainability with the 1 in 10 to 1 in 20 people who have a high baseline, and you can do the math to see how many people have both. According to Bouchard's research, somewhere between 0.1 percent and 0.3 percent of individuals would have the genetic ability to reach the very top of the athletic pyramid.

It's clear that the odds worked out for Paula Radcliffe. The fact that she had a VO_2 max of around 70 while running just twenty-five to thirty miles a week doesn't just suggest that she had an exceptional baseline value. It also suggests that she was highly trainable. Given that her VO_2 max remained at essentially that same level for the remainder of her career, we can guess that even the relatively low training volume she was doing at that point had a very positive effect on her fitness, pushing that value to what turned out to be a stable—and very high—number.

So how did Radcliffe get faster, if by age eighteen she already had basically the same VO_2 max she'd have for her whole career? The biggest key seems to be huge changes in her running economy. Jones conducted tests in which Radcliffe would run on a treadmill at 16 km/hr (that's a pace of six minutes per mile—quite fast for us normal humans, but a steady-tempo run for an elite female athlete). Running economy measurements are in ml/kg/km, basically quantifying how much oxygen it takes for an athlete to cover a certain distance at that 16 km/hr speed. As a rough guide, 200 ml/kg/km is a solid score for a top athlete; lower numbers, which mean the athlete is using less oxygen, are better.

Back in 1992, Radcliffe was tested at 205. By 2003, that had dropped all the way down to 175. So while eleven years of desperately hard training had little to no effect on Radcliffe's VO_2 max, it resulted in a 15 percent increase in her running economy.

What accounted for that increase? Jones isn't exactly sure. "What we can't do with running economy is differentiate between differences in biomechanics and differences in physiology," he says. "You get these subtle changes in running technique over many years that reduce the energy cost of movement. Simultaneously, you've got changes to muscle fiber types, and that's a physiological adaptation that complements what happens biomechanically. So it's very difficult to know exactly what it was, because so many things can change simultaneously."

What's interesting is that Radcliffe didn't look like the picture of a superefficient runner that most coaches imagine in their heads, all flowing movement and relaxed motion. She would sort of grimace when she was running, and seemed to have a lot of muscle tension in her upper body. In fact, even the best coaches can't judge economy by looking at a runner's form. Jack Daniels, a longtime running coach, once taped the form of a group of runners, then asked a set of coaches to rank them in order of economy. No one got it right. Radcliffe's form wasn't pretty, but it was devastatingly effective. The proof was there in the numbers. "None of her world records surprised us," says Jones. "The testing we had done beforehand showed us she was capable."

Expressing Yourself

The advent of the human genome project and the decreasing cost of gene sequencing has led to some pretty optimistic views about how easy it would be for scientists to find the genetic answers to some of the questions that have troubled humanity for generations. We'd find genes that would allow us to cure diseases and extend our lives and health. We'd find genes for good eyesight; we'd find genes that caused high blood pressure or diabetes. For those of us interested in sports, we'd find the genes that allowed for superior aerobic performance, or for the exquisite reaction times necessary to return a tennis serve blasting at us.

Cue the sound of a sad trombone. Because it's nowhere near that easy. There's a big difference between saying that a trait is genetically determined and being able to identify associated genetic differences at the molecular level.

In fact, there are a staggering number of genetic variations involved. A paper in 2010 used the results of many years of what are called genome-wide association studies to try to identify the factors involved in height. In this sort of study, you take a large pool of individuals and then look for single nucleotide polymorphisms (or SNPs, pronounced "snips"). A SNP is a variation in a DNA sequence in which just one nucleotide is different in the genome; a single letter of the genetic code differs between the two samples. If you gather enough people, you can start to associate certain SNPs with the differences between those people, like their height.

That 2010 paper looked at nearly four thousand people, and analyzed 294,831 SNPs that were associated with height variance. After combining all the data, the research team estimated that all those SNPs combined accounted for nearly half of the observed variance. As South African sports scientists Ross Tucker and Malcolm Collins write, "Athletic performance is undoubtedly more complex than height, and if such a large population and almost 300,000 SNPs are able to account for only 45% of the variance in height, then the concept that a single gene, or even a few thousand genes, can explain athletic performance is grossly oversimplified and may ultimately be futile."

Single-candidate gene studies like those looking at ACE or ACTN3 have led to some interesting results and research, but most scientists in the field today are excited about what they call the switch from genetics to genomics, from looking at single genes to data-intensive work like genome-wide association studies.

Bouchard has made the switch to genomics in his own research. After all, he has done research showing that about 47 percent of the difference between high responders to training and low responders is heritable. But finding the exact genetic differences between those two groups that are responsible for that difference is another matter altogether. If you could do so, you would have some way of identifying people who are particularly trainable before they even start. That could be a powerful way to screen athletes early in their careers, essentially giving coaches and trainers a certain amount of understanding of their ultimate aerobic potential. So Bouchard and his research group ran a genome-wide association study to find SNPs associated with those differences in training responses.

The team started with a list of 324,611 SNPs to examine. Using computers to crunch the differences in the high responders' and low responders' DNA, Bouchard and his team whittled the list down to twenty-one different SNPs that were associated with the level of their training response.

If a person had nine or fewer of the identified genetic markers, it turned out that his or her training program had an effect that was far below average. These people were the low responders. But, on average, if a person had nineteen or more markers, he or she was a high responder, gaining almost three times as much aerobic fitness as the low-responding group. Overall, the twenty-one-SNP panel identified about half the variation between high and low responders—extremely close to Bouchard's previous estimates of genetic heritability.

The complexity of our individual responses to exercise and athletic training is staggering. Here's an extreme situation: We usually think of exercise as an unalloyed good—something from which every single person benefits. But the story isn't quite that simple. There are people for whom exercise, cruelly, actually causes an *increase* in risk for chronic conditions like heart disease and diabetes.

Bouchard's lab found that around 10 percent of participants in several exercise studies actually had an adverse reaction to the exercise in some way, from an increase in blood pressure to an increase in insulin levels to a change in blood chemistry that increased the risk of heart attacks. That didn't mean the subjects didn't get some sort of cardiovascular benefit from their workouts, but it does mean that the benefits aren't crystal clear. Unfortunately, we can't know at this point if we're one of those people or not, unless we track our health over time. Bouchard suspects that the reason for these negative outcomes is genetic, and that someday, we might be able to develop a test that screens people for these risks. But that's still a long way off. Until then, we simply have to track markers like insulin and triglycerides, and make sure that the exercise we're doing isn't actually causing them to worsen rather than improve.

What Bouchard's work points out is how radically different the individual response to exercise and training can be. On one end of the spectrum are the supertrainable folks who get massive benefits from physical activity; on the other are people who seem to actually be harmed by it.

Even an athlete's risk of suffering certain injuries seems to have a genetic component. Take a common but devastating injury suffered by athletes—a tear of the anterior cruciate ligament in the knee. An ACL tear can happen because of a huge hit to the knee, an awkward cut on the court, or a bad fall on the ski slopes. But it also seems to strike some athletes in situations that seem totally normal, like just running down the field.

It turns out genome-wide studies have identified SNPs that are associated with increased risk of ACL injury. You can also find SNPs that are linked with other common athletic maladies, like Achilles tendon problems, bone density issues and stress fractures, and deficiencies in vitamins and minerals. We all seem to be more or less likely to have these sorts of problems based on our genetics.

Dr. Stuart Kim leads a lab at Stanford University Medical School that has collated all of these genetic markers into a genetic testing panel that can be used to evaluate a specific athlete's genome. Kim walked me through a hypothetical case involving runners. "One team that we've worked with, seventy-five percent of the runners on the team are injured each year," says Kim. Imagine you were the coach of that team, and through this genetic screening you found that a specific athlete was at increased risk for stress fractures.

"You can then modify that runner's training and diet to try and ameliorate that risk," says Kim. You can make sure that her diet is supporting good bone health. You can check on her biomechanics to make sure they're not causing undue damage. You can look at her shoes, and evaluate their effect on her running. And you can modify her training program so she has enough time to recover and stay healthy."

Elite athletes live on a knife's edge when it comes to training and injury. They push and push and push until their bodies break. By evaluating where the weak links in a particular athlete's body might be, Kim's screening test could give coaches a way to tailor workouts more effectively, and help their athletes go right up to the edge of injury without getting there.

There's an old saying: The most important ability for an athlete is availability. If you're hurt and can't play, you can't win.

Genetics describes the effect of your genome, the sequence of your DNA that codes all the information in your cells. But there's a cellular layer above the genome that scientists are starting to understand as having a huge effect on how our bodies grow, develop, and behave. The scientific term is *epigenetics*. The concept here is that there are changes in how our genes are expressed—some of which are heritable—that happen even without any change to the DNA sequence itself. To think of it another way, our genome is like the hardware in a computer, while our epigenome is a piece of software that controls the operation of that computer.

That's how one genome can create the thousands of different kinds of cells in our body—that genetic material is expressed differently by each cell through epigenetics. But one important thing about your epigenome is that it's not stable or static. The way our genes are activated and controlled can be influenced by factors like diet and hormones. For athletes, there's even some suggestion that the use of performance-enhancing drugs could alter an athlete's epigenome, meaning that there might be some effects from doping that persist even after an athlete stops taking the drugs (more on that in Chapter 11).

With as little as we know about the role of specific genes in sporting performance, we know even less about the role of epigenetics. Could diet change how certain genes are expressed? There's a pretty good chance that it could. But the complex interaction of all these factors shows just how far we have to go before we will fully understand the role that genetics plays in athletic greatness.

Cracking the Code

The HERITAGE study group's data has powered Bouchard's research for nearly twenty years. But as he and I talk about his work, I can tell that he's starting to feel the limits of the cohort involved, especially the fact that there weren't elite athletes in the study—everyone who participated started as a completely inactive person, so that the effects of training could be determined. Now, says Bouchard, the basic science is mature enough for what he calls a "grand project" to identify the genetic factors that influence the development of elite athletes.

"You'd have world-class endurance athletes, and world-class speed and power athletes, and world-class sprinters," he says. "You'd have a large sample of those, and compare them to a very large group of sedentary people. You'd look for the genomic differences, the proteomics, the metabolomic differences as well. So that you get a good picture of the underlying biological differences between these world-class, elite samples and people who are so far away from those performances."

As he describes the study, Bouchard starts to talk faster and faster. It's clear that he's thought about this dream project for years.

"I'd start with sequencing the whole genome of these people, and then identify genes. For a subsample, I'd like to take blood cells, small muscle biopsies, and do a transcription profile. What are the genes expressed in the athletes at the high level that aren't expressed in the sedentary people? What are the differences between the endurance athletes and the sprinters?"

After identifying the target genes more common in elites, Bouchard would move on to a bigger version of HERITAGE. "I'd take a large cohort, probably a few thousand subjects who are sedentary, and endurance-train them," he says. "The same with another group and muscle power and strength-train them. Same with another group, and sprint-train them. So that we understand the true range of human variation and the ability to respond."

But so far, the money isn't there to do the research. Bouchard has tried to get funding for the endurance portion of his grand plan, and pegs the cost at about $125 million (roughly the cost of the five-year contract Josh Hamilton got in 2012 to play outfield for the Los Angeles Angels). With the billions and billions of dollars circulating in the sports world, with teams spending hundreds of millions on individual players, Bouchard is surprised that the money can't be found to answer these fundamental questions.

"If the USOC [United States Olympic Committee] had vision, they would fund these studies," he says. "If the IOC had vision, they would fund them internationally, because these aren't studies that need to be replicated two or three times. With a few hundred million dollars they would provide enough information to allow the biomedical community to work on this for decades."

Or, of course, a team or a country could decide to fund the work, and keep the results for themselves. If you could identify potential elite athletes before they even start to train, you would have a huge competitive advantage, one worth much more than a few hundred million dollars. You'd have cracked the code.

"What are the true predictors?" says Bouchard. "Can we, based on the profile in the sedentary state, identify those who are likely to reach the elite level, versus those who will have an average response, or those who are destined to be poor or nonresponders? That would be the dream. I'm ready to go!"

3

EATING TO WIN

Supplements, Snake Oil, and Everything in Between

In 1965, an assistant football coach at the University of Florida asked Dr. Robert Cade, a kidney disease specialist, a fairly simple question. Why did players lose so much weight during games—up to 18 pounds in some cases—and why did they urinate so little? After thinking about the question for a moment, Cade supposed that it was because the players were sweating so much that there basically weren't any fluids left in their bodies to urinate.

Did that really matter? Cade started to work with the Florida football team to find out, testing players on the freshman team during practice to see what was happening to their blood chemistry. Cade found that they had low blood volume from dehydration, out-of-balance electrolytes, and low blood sugar. With his research team, Cade concocted a drink: salt to help with electrolyte replacement, sugar to keep blood sugar levels up, and, of course, water for hydration. The first batch tasted horrible, but Cade's wife suggested adding lemon juice—and Gatorade, named for the university's reptilian mascot, was born.

After a first test in an intrasquad scrimmage, the Florida head coach asked Cade and his team to make up enough of the magic elixir to fuel the

Gators the next day as they took on heavily favored Louisiana State. Swilling their secret weapon throughout a 102-degree day, the Gators came from behind to win 14-7.

Today, the shelves at your local grocery store fairly groan under the weight of the sports drinks housed there, all spawned from that first homemade batch in Gainesville, Florida. There's regular Gatorade, low-calorie Gatorade, endurance formulas, versions made with all-natural ingredients. There's Powerade, Accelerade, Vitaminwater, Propel, Powerbar Endurance, all in a rainbow of iridescent, fruitish flavors.

Then there are all the other products meant to fuel athletes before, during, and after training and competition. There are hundreds of energy bars, with differing formulations that mix and match protein and carbohydrates. There are gels in flavors from chocolate to lemon to blueberry cheesecake, meant to give a quick hit of carbs in an easily digestible form. And if you get sick of those, you can get something chewy, like a highly engineered gummy snack that's aimed at the same fueling goals. There are Shots and Sharkies; Gu and Blasts.

This is a big business—in 2013, the market research firm Packaged Facts estimated that sports drinks and nutrition bars was a $10 billion-a-year business in the United States alone—and all this from one homemade drink. But Gatorade didn't just herald a new industry; it also created an idea: that athletes need special fuel and nutrition in order to perform at an optimum level.

In some ways, that's completely true. The energy demands of some sports are outrageous when compared to the caloric intake that even a dedicated recreational athlete requires. But in other ways, we seem to have lost sight of that difference; the fueling requirements of an elite athlete have very little to do with what most of us need. In fact, there's a good chance that you and I don't really need to down that bottle of Whatever-Ade.

The typical athlete, without using any special nutritional techniques, has enough carbohydrates in his body to fuel roughly three hours of endurance exercise at around 70 to 80 percent of his maximum effort. This seems to run counter to the marketing messages that tell us we need to constantly replenish our body's fuel during exercise. Just think about the typical person who hits the gym for thirty minutes on the elliptical machine: She might only burn about 350 calories, and then put 200 calories in carbs right back in by drinking a thirty-two-ounce sports drink. There's no increased performance, and if you're looking to lose weight, you're not making much progress at all.

In fact, Asker Jeukendrup, the sports scientist who now heads up Gatorade's own research lab, has published guidelines on carbohydrate intake that indicate that if you're exercising for less than seventy-five minutes, you probably don't need any carbohydrate intake at all for optimal performance. Your body already has plenty of fuel for these shorter efforts without any sort of bars or gels or drinks.

But now let's look at what a rider on the Tour de France requires. He has extravagant caloric burdens to deal with during the six hours he spends on the bike each day. Just imagine living a day as a Tour rider—not just on the bike, but at the table. You begin with a light breakfast: half a pound of pasta, a pastry, some muesli, a couple of pieces of fruit, coffee, OJ, water. Before the race starts, you'll have another half pound of pasta or some rice, and start slamming down water to hydrate yourself. At this point, your daily intake is approaching two thousand calories, and really, you're just warming up.

Because here comes the race itself, when the real fueling starts. Each hour in the race, riders eat about three hundred calories' worth of energy bars, gels, and chews—on a long, six-hour day, that's eighteen hundred calories. Teams are starting to augment these highly engineered foods with more recognizable snacks like small sandwiches, sticky rice cakes, or even boiled potatoes with olive oil and Parmesan cheese (riders appreciate the variety and the more recognizable nature of these foods). And then there's the drinking—roughly one twenty-ounce water bottle filled with either a sports drink or plain water every half hour. Near the end of a day's racing, you might get a little treat from the team car—a can of Coke for an extra sugar and caffeine boost.

At the day's end, it's time for recovery. A protein-rich recovery drink helps muscles rebuild and repair themselves, and more rice or pasta provides carbs for energy. More water, of course, maybe a little candy. Dinner is more grains (pasta or rice), some lean meat like chicken breast, lots of veggies, some cheese. Dessert might be yogurt or a fruit crisp. Oh, and more water, another quart at least. By the time you're having a bedtime snack—maybe a small piece of chocolate, some more protein drink—you're likely to have downed in the neighborhood of nine thousand calories for the day, roughly triple what a normal, healthy, active man might eat. But you won't gain weight—that's how much energy you are burning each day as you chase the yellow jersey.

Athletes eat; elite endurance athletes eat *a lot*. During the Beijing Olympics, stories started to circulate that swimmer Michael Phelps, who was in the midst of the greatest individual performance at any Olympic Games, was putting away around twelve thousand calories a day. Phelps later said that number was exaggerated, but it's likely that he was eating in the range of eight thousand calories. The world that elite athletes live in is very different from the one most of us inhabit, and nutrition might be the prime example of that phenomenon. Even if you wanted (for some ungodly reason) to eat nine thousand calories tomorrow, you'd be hard-pressed to do so without some pretty gnarly gastric distress.

A Tour de France rider needs all those sports drinks and gels—it's not easy to get that many calories into a body. An elite marathoner needs to be replacing carbs, because he's likely to burn through his body's built-in energy supply about ninety minutes into the race unless he eats. And running at those high speeds can result in a serious amount of sweat: up to three liters an hour in one study of a 2:06 marathoner. So he'll have to keep an eye on his hydration, although we now know that most elite runners actually finish their races slightly dehydrated.

There's a lot of marketing—and hype—around sports nutrition. These products have been an amazing boon to the very best athletes in the world, who are taxing their bodies and their energy systems to the absolute limit. But to draw a line from that experience to the typical gym rat? That's probably a mistake. It's important to be realistic about how much energy you're burning, and avoid simply pouring in carbs and calories and fluids that you don't really need.

There are a couple of easy guidelines to follow. The first concerns the dangers of dehydration. For years, many athletes have been told that dehydration is a constant hazard, and that they need to get on a set schedule of replacing fluids. While it's true that dehydration can be dangerous, it's not a supercommon issue for most people, and we now know that elite athletes in events like the marathon will lose a certain percentage of their body weight in sweat without performance problems.

The good thing is that every human has a very sophisticated internal system that lets us know when we need more fluids without having to do math about our body weight and the intensity of our exercise: We get thirsty. The best guideline is the simplest one—drink when you're thirsty, and don't worry about drinking if you're not.

Now, when you do drink, should it be a sports drink? It totally depends on the sort of exercise you're doing. During what sports scientists call "brief exercise," of less than forty-five minutes, guidelines from some of the top sports nutritionists suggest that you don't need any carbs whatsoever. In the space from forty-five to seventy-five minutes, a little bit of carbs can be helpful, mostly due to the effects that carbs seem to have when they hit your mouth, even before you swallow them. There's a strong line of research that shows that we have sensors in our mouths that detect the presence of carbs, and that even just rinsing your mouth with carbs has an effect on your brain that can increase your performance. More on this in Chapter 8.

When the events get longer, your energy requirements get higher. In endurance exercise as well as "stop and start" sports, like soccer and basketball, that last from an hour to two and a half hours, the recommendation is to consume between 30 and 60 grams of carbs per hour (a twenty-ounce bottle of Gatorade has 34 grams of carbs).

And you don't have to use a commercial sports drink—the U.S. Olympic Committee's website highlights a recipe from sports nutritionist Nancy

Clark. Take ¼ cup sugar and ¼ teaspoon salt, and dissolve them in ¼ cup hot water. Add ¼ cup orange juice, 2 tablespoons lemon juice, and 3½ cups cold water. You've just made a sports drink with basically the same nutritional profile as the versions in the grocery store.

When your efforts really go the distance—more than two and a half or three hours—fueling becomes a bigger issue. For elite athletes, the recommendations are to consume up to 90 grams of carbs an hour, the sort of intake that Tour de France riders strive for. Most recreational athletes aren't operating at that level of intensity, but if you're out for a very long run or bike ride, you'll probably need more calories than you can get just from a sports drink. That's where products like gels and blocks come into play. These products tend to have between 20 and 30 grams of carbohydrates per pack; eating one or two of them an hour in addition to a sports drink will provide most of us the energy we need to perform our best.

Pills and Powders

Beyond the components of a healthy diet, athletes are faced with a dizzying array of potential nutritional supplements. If you thought sports nutrition was a big deal, just look at supplements, which have exploded into an industry estimated in 2013 at $30 billion in the U.S. alone. Go into any GNC or Vitamin Shoppe and you'll see shelves crammed with thousands of pills and powders, all with the theoretical possibility of increasing your athletic performance. It's a vast land of confusion, rumor, bad science, belief effects, and in the end, some things that actually do help athletes perform at a higher level.

How many supplements are out there? The *British Journal of Sports Medicine* set out a few years ago to produce a comprehensive A-to-Z series that gave athletes and coaches solid scientific advice on the effectiveness and safety of the most common compounds that were of interest in the sporting world. It took more than four years to complete the survey, and there were forty-eight separate articles in the journal, covering more than a hundred supplements and other products. That's a lot of pills, powders, and compounds.

For most of us, there are a lot of things to do that would be much more useful and effective at improving performance than designing an elaborate supplementation program—train more, for instance. Using many of these pills and protocols takes a lot of effort—and includes some risk—for a pretty small gain. But at the very top level of sports, even 1 percent improvement is a huge difference, so the time and energy spent on supplements by those athletes is well worth it.

Different sports and countries take diverse approaches when it comes to navigating this thicket of choices. One of the most transparent is the supplement program that has been set up in Australia under the supervision of the Australian Institute of Sport. It's a unique window into what some of the world's top athletes are actually using in training and competition.

The AIS has set up a website that classifies supplements under four groups:

Group A: Compounds for which there's good evidence of a benefit, and which the AIS will provide to its athletes as needed.

Group B: Things that show some evidence of a benefit, which will be provided in a research-based protocol.

Group C: Supplements that have no evidence of meaningful effects, which aren't recommended or provided to athletes by the AIS.

Group D: Banned substances, or compounds that have an especially high risk of contamination.

Many of the items in Group A are things that we usually don't even think of as supplements in the traditional sense: sports drinks, energy bars, protein shakes, gels. There are also multivitamins on the list, as well as three specific vitamin and mineral supplements: vitamin D, calcium, and iron.

Vitamin D is synthesized through the exposure of skin to sunlight, but athletes who spend most of their time inside can suffer from chronically low levels. (Imagine that you're a gymnast who spends six hours a day training indoors, or the typical office worker who spends most of the day inside.) As recent research has shown that a lack of vitamin D not only affects bone health but can also cause inflammation and reduce the effectiveness of muscles, many elite athletes rely on blood testing to check their vitamin D levels, and use diet and supplements to keep it in the optimal range. Much the same applies to calcium and iron, both of which are crucial for athletic performance—calcium for bone health and strength, and iron for blood cell development and regeneration. Simple blood tests can help identify chronic deficiencies in these nutrients, and they can be supplemented.

That leaves us with the last three supplements in the AIS Group A. One is the most common performance-enhancing compound in the world, one is more often thought of as an ingredient in cookies, and one needs to be rescued from its bad reputation as being related to steroids.

Give Us Our Daily Jolt

There are few sports as tied up in tradition as European road cycling. For decades, riders have followed a maze of rules and practices that range from the puzzling to the bizarre. Air-conditioning is frowned upon. Riders will train in long sleeves and full tights even in warm weather, to avoid "catching a cold." No shaving your legs the day of a race, because it will "steal energy." No driving with the windows down. No eating ice cream during competition.

But there's one cycling tradition that actually can lead to increased performance. The classic meeting place for a road ride in Europe is the café, where a rider will down an espresso or two before heading out for a spin. Beyond just being a pleasant way to ease into a day's training effort, those two shots are a nice dose of one of the most effective performance aids around—our trusted friend, 1,3,7-trimethylxanthine, which you might know by its less formal moniker: caffeine.

How effective is caffeine? So effective that having too much in your system was classified as a doping violation under International Olympic Committee rules from 1980 to 2003. The rules were designed to allow for the sort of levels that someone might have from drinking his daily coffee or tea, but to keep athletes from using high dosages to gain an advantage. It turns out that when it comes to athletic performance, there's little dose-response effect for caffeine—taking more doesn't result in a larger bene-fit—so the drug was removed from the banned list in 2004.

The good news isn't just that it's now legal in competition but also that caffeine is a really effective substance for a broad variety of athletes. As a recent review in the *British Journal of Sports Medicine* puts it:

> Caffeine supplementation is likely to be beneficial across a range of sports including endurance events, "stop and go" events (e.g., team and racquet sports) and sports involving sustained high-intensity activity lasting from 1–60 min (e.g., swimming, rowing, middle and distance running races). The direct effects on single events involving strength and power such as lifts, throws and sprints are unclear.

There aren't many things as easy to use as caffeine that have benefits across such a wide variety of activities. Most of the time, a supplement will be helpful to endurance athletes and not power athletes, or vice versa. But caffeine seems to be an effective aid to all sorts of athletes, from the Tour de France rider to the 1,500 meter runner (not to mention those of us just sprinting to catch the train).

Part of what makes caffeine so interesting is that no one is completely certain how and why it's so effective. The drug has effects directly on mus-cle contraction and in the release of adrenaline in the body, but the most popular theory at the moment is that caffeine provides a large part of its benefit by altering the central nervous system's perception of effort and fatigue. That is to say, caffeine tricks our brains into feeling less fatigue, so we can perform more effectively.

For years, some athletes avoided caffeine because of fears that the drug's supposed diuretic effect might lead to dehydration during exercise. But recent research has shown that for athletes (and others) who regularly use it, the most typical doses of caffeine have a very minor effect on their hydration status.

In the past, scientists thought that athletes needed relatively large doses of caffeine, on the order of 6 milligrams per kilogram of body weight, to see performance benefits. But now the AIS says the best practice for using caffeine for athletic performance is less extreme—around 3 mg/kg. For a 175-pound man, that's about 240 mg of caffeine, or roughly what you'd find in two cups of coffee. More than that doesn't offer any increased benefit.

An Antacid for Your Muscles

Go ahead, open up your pantry—you'll see it there. Or maybe you keep a box in your refrigerator, with the top popped to help absorb odors. But somewhere in your house, I'm betting that you have a supply of one of the hottest legal athletic performance supplements in the world: baking soda. We'll give the Arm & Hammer its scientific due and call it sodium bicarbonate. But the white powder in that box has proven to be a big tool in the performance nutrition arsenal.

Sodium bicarbonate operates as what exercise physiologists call an "extracellular buffer." Generally, for a resting human being, blood pH is about 7.4, and the pH in human muscle is typically 7.0. As we exercise, our blood pH can fall to 7.1, and in muscles, it can get down to 6.8. That's because hydrogen ions are released as part of the reaction that produces energy in our muscles, which makes our blood and muscles more acidic. Acidic muscles feel more fatigued than ones that aren't, so anything you can do to prevent acidosis is good. During exercise, the hydrogen ions have to be transported out of the cell. The faster you can get those ions out of there, the higher the pH will stay, and the less fatigued your muscles will feel.

Sodium bicarbonate raises the pH in the blood, which helps the body clear hydrogen ions more quickly from muscles. (The greater the difference in pH inside and outside the cell, the faster the ions are moved from the cell to the blood. The body is trying to balance the pH across the two.)

Make sense? For anyone who might have been napping in chemistry class, think of sodium bicarbonate as an antacid for your muscles that allows them to function more effectively.

There are some important caveats, though. The first is that the kinds of exercise that are reliant on these energy systems tend to be shorter, more intense efforts. Endurance athletes primarily rely on different systems, and there's not a lot that sodium bicarbonate can do for them. The research indicates that the athletic event needs to be between one and seven minutes long for sodium bicarbonate to really have an effect on an individual's performance. Of course, this is a pretty common time frame for sports: Middle-distance track events, rowing, swimming, and track cycling all feature events of this duration. In those events, the benefits can be relatively impressive. A 2011 meta-analysis of bicarbonate supplementation found that athletes in short-duration events could expect about a 2 percent increase in performance when ingesting the standard dose of 0.3 grams of sodium bicarbonate per kilogram of body weight.

Oh, that's the other problem: taking it. For a 175-pound athlete, the recommended dose works out to 24 grams of baking soda, which doesn't sound like much on paper. But when you translate that into the thirty or so capsules of sodium bicarbonate you'd have to take, or the ¼ cup that it measures out to in my kitchen, it might not come as a surprise that what scientists delicately refer to as "gastrointestinal side effects" can be a huge problem with bicarbonate supplementation. In fact, a research team from the AIS reports that the potential problems, including nausea, stomach pain, diarrhea, and vomiting, have kept many athletes who could benefit from sodium bicarbonate from using it at all. Baking soda can help your performance, but those difficulties in taking it mean that, unlike caffeine, it's likely to remain a tool for elite athletes and iron-stomached amateurs. If you're determined to give it a shot, try combining the bicarbonate with a small, carbohydrate-dense meal between 120 and 150 minutes before exercise, and spread out the doses to minimize your discomfort.

The One with the PR Problem

In 1998, as Mark McGwire and Sammy Sosa were chasing the single-season home run record, creatine suddenly became part of the media story. An AP reporter wrote that McGwire used the supplement, along with a product called androstenedione. At that point, androstenedione was actually banned by the NFL and the International Olympic Committee, and it's banned by most sports today. We've since learned that baseball had a huge problem with anabolic steroids in that era, and that's just what andro is.

Creatine and andro were kind of lumped together in those discussions, leading a lot of people to view creatine as being a type of steroid, which it is not. Creatine is a compound that our bodies create naturally, mostly in the liver, by combining three amino acids. It's used by the body to help regulate the energy cycle inside muscle cells. Muscles power their contractions using a molecule called adenosine triphosphate (ATP); creatine helps muscles regenerate ATP stores more quickly during intense efforts.

This leads to a host of positive effects for the athlete. The most common use of creatine is in conjunction with weight training—the more rapid regeneration of ATP between sets of lifting allows an athlete to train at a higher intensity than he or she could without the creatine. In one study, lifters taking creatine improved as much as 14 percent more than lifters who weren't taking the supplement while on the same training protocol.

But weight training isn't the only sort of training that involves high-intensity bursts of effort with short recovery periods. Creatine has been shown to have benefits for any sort of sport that involves repeated efforts of thirty seconds or less. Tennis, for example, involves just this sort of activity pattern: During the point there is maximal effort, followed by a chance to recover briefly. The AIS guidelines suggest that creatine is useful for that type of situation, and for players in soccer, basketball, and football as well—what they call sports with "intermittent work patterns."

There are even some suggestions that creatine can help with longer, aerobic-type endurance exercise, by exerting a positive influence on other factors that help performance. Creatine seems to help increase the glycogen stores in muscles that are important to endurance exercise, as well as possibly reducing oxygen consumption in submaximal efforts. One caveat for endurance athletes: Creatine can lead to water retention, which can cause unwanted weight gains.

Simply put: When looking at all the science, creatine works. There have been concerns in the past about creatine supplementation and kidney and liver function, but at the usual level of supplementation used by athletes, no one has been able to establish any health risks. Given all that I've learned about creatine through researching this book, it's the one supplement that I've started using myself to help maximize the benefits of the weight training I do in the gym. I'm following the same recommendation that the AIS gives to its athletes—three grams of creatine monohydrate a day—to maintain the optimal level in my muscles.

The Two B's

The supplements in Group B of the AIS list, the ones that have some evidence of a positive effect without definitive proof, really show where scientists are pushing the frontiers of the field. There's carnitine, a compound found in red meat that might help muscular function; HMB, a metabolite of an amino acid that might have similar effects; quercetin, a plant pigment that may work as an anti-inflammatory. None of these have strong research backing them as yet—they're firmly in the "maybe" column.

But two of the items in this group have rocketed to prominence in the past several years, as more research piles up that seems to demonstrate that they provide clear benefits to athletes (for the AIS to promote a supplement from Group B to Group A, the science has to be basically beyond argument). The first is the latest craze in sports drinks—not something citrusy from one of the big sports labs; not even chocolate milk, which has been shown in study after study to be a great, low-cost drink for recovery after a hard workout.

No, today's hottest sports drink is deep red and frothy, and tastes a little bit like dirt. Drink the recommended dosage, and you may find that your urine and feces become pink from taking it. But you also might find that you're faster in your races. Ladies and gentlemen, I give you beet juice.

The key researcher into beet juice's effect on athletes is Andy Jones, whom we met earlier through his work with marathoner Paula Radcliffe. How much has Jones, a professor at the University of Exeter, in the UK, become associated with the beverage? So much so that you can find him on Twitter under the username of @andybeetroot. Jones's group has published a study that seems to show a nearly 3 percent gain for athletes involved in events that last between five and thirty minutes. The results are more ambiguous for longer events—the Exeter researchers found that while there were small performance increases for cyclists in a fifty-mile time trial when using beet juice, they weren't quite large enough to be statistically significant (although that doesn't mean that they aren't real, or that they should be dismissed out of hand).

The compound responsible for these effects is nitrate. The body transforms nitrate into nitrite, and then into nitric oxide. According to Jones, nitric oxide has two major effects on an athlete. "The first is that it causes blood vessels to dilate, so you can provide more blood through them," he says. "Simultaneously, it seems to make the mitochondria more efficient, so they are able to create the same energy while consuming less oxygen. So you really have two things happening. Lower oxygen cost because the mitochondria are more efficient, and then you have a higher oxygen supply—in terms of performance, that's a pretty good combination."

We're commonly told that nitrates and nitrites are potentially dangerous, and that we should limit our consumption of them. The fear is that inside the body, nitrates and nitrites can combine with meat proteins to form compounds known as nitrosamines. There is evidence that these compounds are carcinogenic, which is the reason that most health organizations advise that we limit our intake of cured meats like bacon and hot dogs, which use sodium nitrite in the curing process.

But Jones and his team have shown that we're still very early in our understanding of what nitrates and nitrites do in our bodies, especially when it comes to athletic performance. As opposed to cured meat, beet juice contains nitrate, not nitrite, and there's no protein that could lead to the formation of nitrosamines. The early results have been good enough that Jones tells me just about every top nation at the 2012 Olympics was using beet juice with its athletes. "It was actually pretty difficult to buy beet juice within ten miles of London," he says.

Up until now, the studies have been focused on the drink's acute effects on performance, like drinking it right before a competition. But Jones and his team are starting to study longer-term usage in a training setting. It could be that drinking beet juice consistently would allow athletes to train harder, leading to better adaptations, but it's hard to say until the research is completed.

One other interesting result of Jones's research is that beet juice seems to be a more effective ergogenic aid for regular athletes than it is for elite performers. "If you think of the things beet juice helps with, like blood flow and mitochondrial function, in elite athletes, those abilities are pretty well developed. So there, you do have an issue of diminishing returns—any ergogenic aid might have a smaller benefit in the elite. But even if the benefit is just 0.1 of a percent, it's probably worth trying." This is one case in which regular folks like you and me might get more out of beet juice than an Olympian.

The other hot new supplement is beta-alanine, an amino acid that's found only in animal proteins, particularly in fish and white meat like chicken breast. When beta-alanine combines in our bodies with another amino acid (histidine), it forms a compound called carnosine, which is found in large amounts inside our muscles.

Carnosine is another buffer against acid buildup in muscles. We saw that baking soda could act as an extracellular buffer for hydrogen ions by raising the pH of the blood outside the muscle cell. Carnosine does the same thing, but it's an intracellular buffer, helping to reduce the harmful effects of acidosis inside the muscle cell.

Most of us don't have as much carnosine in our muscles as we could because we don't have enough beta-alanine. There doesn't seem to be an upper threshold on how much carnosine we can make, so increasing beta-alanine seems to be very effective. The interest in beta-alanine as a sports-performance supplement started off in 2006 when a research group in the UK showed that taking it could raise the level of carnosine in muscles by 40 to 60 percent. If supplementation could cause such a large increase in the amount of carnosine in muscle tissues, would there be an increase in athletic performance due to better buffering?

The answer so far is a cautious yes. In a 2012 meta-analysis of research on beta-alanine, the average performance increase for subjects taking the supplement was 2.85 percent. Beta-alanine was effective in the same sort of medium-duration efforts as sodium bicarbonate, from one to four minutes or so. Any shorter and the buffering effect isn't as necessary. Any longer and the results start to become more ambiguous.

So, if one buffer is good, are two buffers better? The fact that beta-alanine and sodium bicarbonate seem to work in the same way, but in different cellular spaces, seems to suggest that there might be a synergistic effect between them. Or as Asker Jeukendrup, the Gatorade scientist, writes: "One could term the pH buffers inside the muscle cells (such as carnosine) as the first line of defense and the blood buffers as the second line of defense."

So far, there have only been a few studies in which the two supplements have been combined—and it seems like there's about a 70 percent chance that the two are more effective together than separately. That's not quite enough for a scientist to proclaim that there's a strong effect in a journal, but you have to remember, elite athletes aren't publishing research papers; they're trying to win races. As Louise Burke, of the AIS, puts it, "I'll take that small possibility" when it comes to top-level performance. For the rest of us, it might be a little early to be bringing beta-alanine into a central place in the supplement world, although it's worth keeping an eye on.

Training the Gut

Triathletes sometimes talk about nutrition being the "fourth event" in a triathlon—like the Tour de France, the energy demands of an Ironman race are so extreme that it can be physically difficult to take in enough fuel and process it efficiently enough to allow you to complete the race. An elite racer might burn upwards of ten thousand calories over the course of the event, and running out of fuel for the muscles is one of the biggest pitfalls triathletes face. A key concept that's emerged for these athletes is the idea of "training the gut" to deal with the onslaught of food.

We don't usually think of training ourselves to be able to eat, but like our cardiovascular system and muscles, our stomachs and intestines can adapt when they're given a novel stimulus. This insight has led to some interesting ideas on the frontiers of sports nutrition. For instance, we know that the body responds most effectively to varying levels of training intensity—hard one day, less load the next to allow for recovery, with both short-term and long-term cycles to continually challenge the body in new ways.

You could combine those varied workouts with varied levels of fueling and nutritional support. For instance, you could undertake a hard work-out with a lower-than-usual level of carbohydrates to fuel the muscles. Would that lead to an adaptation from the body so that it would become more efficient in fueling itself? And would that increased efficiency then boost performance when the muscles were given a normal amount of fuel to burn?

These are the ideas driving research into "train low, race high" nutrition. You can think of it as an analog to altitude training, in which the athletes put themselves in a resource-constrained environment (either oxygen or carbohydrate) during training to drive physiological changes that offer a benefit when that resource is restored.

The specific mechanism for train low, race high nutrition centers on the role of glycogen in the muscles, and how it interacts with proteins.

Glycogen is derived from carbohydrates, and it's the main fuel for muscles as they work. But when there is a lower concentration of glycogen in the muscles, the transcription of several genes that are involved in training responses is enhanced (some of those genes are activated by transcription factors that bind with glycogen, and when glycogen is low, those transcription factors can interact with different proteins). That leads to a greater response to the same training stimulus.

Danish researchers did one of the first studies to show this relationship in 2005. It was an ingeniously designed experiment: Seven subjects trained to increase their performance on a leg-extension exercise. They trained one of their legs for an hour each day. The other leg trained in two separate one-hour sessions every other day. The total amount of training was identical, but the fuel conditions were not.

The key here is the timing of the second session. It occurred after the normal session for the leg that was trained every day, which meant that the body's glycogen stores had been depleted by the earlier efforts. So, basically, one leg trained in a high-glycogen state, and the other in a lower-glycogen state.

At the end of the ten-week period, the low-glycogen leg was able to perform much longer before becoming exhausted—nearly twice as long as the high-glycogen leg. The low-glycogen leg also showed a much improved resting concentration of glycogen. It had adapted to the low fuel conditions by learning to store fuel more efficiently. Also, the levels of citrate synthase, a key enzyme that helps regulate the muscle fuel cycle, were increased more in the low-glycogen leg than in the high-glycogen leg.

This was a powerful finding, the sort of thing that can cause sports scientists and nutritionists to freak out a little bit. It seemed that you could use different fuel conditions to hack an athlete's training and get more benefit from the same level of effort. But going from a lab setting to the real world is always tricky, and in this case, it was harder than usual. Follow-ups that built on the first studies have, in fact, found that there's not a clear line between those molecular adaptations and increased performance in the real world. Why don't these enhancements at the muscular level increase a competitor's performance? In a review of this train-low re-search, Louise Burke writes:

> Reasons for this apparent disconnect include the brevity of the study period, the possibility that performance is not reliant or quantitatively linked to the markers that have been measured, our failure to measure other counterproductive outcomes and our focus on the muscular contribution to performance while ignoring the brain and central nervous system. Most impor-tantly, we may again be simply unable to measure performance well enough to detect changes that would be significant in the world of sport.

It only takes a fraction of a percentage change to be meaningful at the elite level, and it's really hard to detect changes that small. That's why, de-spite the ambiguity of the results, researchers and coaches continue to ex-plore train-low fueling methods—the lab results are strong enough, and the possibility of improvement so enticing, that unless there's evidence of harm, it's worth it to try.

In the Zone

Nutrition is such a fraught subject—it seems like everyone you encounter in your life has a pet theory, or a diet, or some deeply held perspective on what you put into your body. I've got friends who are on high-fat, low-carb diets; gluten-free; vegans and vegetarians; raw food advocates; ob-sessed with fermented foods. We all seem to be looking for something to help us feel better and perform our best, and as a species, we seem to have a limitless credulity for new ways to do so.

Our diets, and the way our bodies deal with the food and nutrients we give it, are intensely personal and individual. I think part of why we see so many fads and swings in the popular literature about diet is that we forget this fact, and look for one way of eating that works for everyone in the world. And there simply isn't one. Like so many factors in performance, N=1 when it comes to diet.

That's why I'd been looking forward to getting the e-mail that just popped up in my in-box. Its subject line seems to mirror my excitement: "Your InsideTracker Results Available!" Inside are the results of a blood test I had a week ago in a crowded lab in downtown San Francisco, where a phlebotomist drained six vials of my blood and then sent them out for various tests. The results were sent to InsideTracker, a company started by doctors, scientists, nutritionists, and exercise physiologists from Harvard, MIT, and Tufts. The Cambridge, Massachusetts–based startup promised to analyze my blood for twenty different biomarkers and then advise me on how they can be optimized for athletic performance.

You'll notice the phrase "optimized zone" when I'm talking about the levels in this testing. Instead of using the normal ranges used in most lab testing, InsideTracker uses these zones, which are calculated in two ways. First, the company collates the scientific literature for these markers, with a special emphasis on studies that reflect performance. And then there are the 150,000 or so other users of the system, including professional athletes from all of the major American team sports and many other individual elite athletes. "We're constantly refining our database to understand the optimized zone for athletes," says Gil Blander, the company's founder and chief science officer. It's a feedback loop into the system.

Part of the InsideTracker results are some of the more standard blood work—so, for instance, my total cholesterol is good at 159, but my HDL is a little low and my LDL is a little high. I probably could eat less red meat and more fish. White blood cell count, fasting glucose—both look solid.

Then there's my vitamin D levels, which are low, about half of the optimized zone that InsideTracker recommends. This isn't a huge surprise, as up to 60 percent of the people whom the company tests show low vitamin D levels, according to Blander. The site includes recommendations on how I can fix this, including suggestions to eat more eggs, eels, mackerel, and herring. Getting out in the sunshine more often would help, but that's always a struggle with a desk job that keeps me indoors most of the time. Or, Blander suggests, I could consider a supplement, which is probably more likely than a radical increase in my mackerel intake. My calcium is also a little low, although an increase in Vitamin D might help me absorb calcium more effectively.

In a call with Blander after I get my results, I ask him about high-fat, low-carb diets for athletes. He says that again, it's a very individual thing. "I see from your numbers that you're dealing with carbs very well," he says, "but you have a small issue with fat. For you, high carbohydrate is good. High fat isn't. But for some of our customers, we see them eat high fat, and do well with that. We're not telling you what to do, but we can tell you how it works for your body. We can see it in the biochemistry."

Then I come to one result that rattles me: My testosterone is below the optimized zone. I can't imagine that there's a man on the planet who is happy to hear that he has low testosterone, especially one who's in his forties and is starting to think about the fact that he's getting older. The first thing I do is head to the National Institutes of Health website and see what the normal range is, as opposed to the InsideTracker optimized zone. I find that while I'm below the median, I'm still within the normal range, clinically speaking.

But for an athlete, I'm low. Testosterone, especially for male athletes, is a big deal. That's why it's a banned drug in the antidoping rules—it's really powerful. It helps build muscle and increase strength, energy, and aggression. For an athlete, all of those are pretty darn useful.

Blander says that 20 percent of InsideTracker's customers come up with testosterone that's outside the optimized zone. One of those customers was Pittsburgh Pirates relief pitcher Mark Melancon, who started working with the company in October 2012. "We saw some results with Mark in biomarkers that are related to performance, like testosterone and creatine kinase, and markers that show inflammation," says Blander. "When you exercise too much, you see a dip in testosterone, and an increase in creatine kinase, which shows muscular damage. So we started working on his diet to help those markers and inflammation."

In 2012, pitching for the Red Sox, Melancon had a poor 6.20 ERA in forty-one appearances; in 2013 with the Pirates, he was one of the most dominant relievers in the game, posting a 1.39 ERA in seventy-two appearances, and making his first all-star team. "I can't say it's all because of the work with us," says Blander, "but we're glad to be part of his team."

So, perhaps I shouldn't be so stressed out about this—especially since stress can actually reduce your testosterone level, along with a lack of sleep. Blander tells me not to be too concerned, since the value is close to the optimized zone. "Think of it as a check engine light for your body," he says.

Blander suggests a couple of ways I can improve my levels. First, try to get more sleep, and also get more exercise (the irony of writing a book about sports is that you have much less time to participate in them!). Also, eating foods that support testosterone, like avocados, cashews, and eggs, can help.

At places like the U.S. Olympic Training Center, these sorts of blood tests are common—most athletes are tested at least annually, if not more often. Companies like InsideTracker have brought the tests to the masses, or at least the masses who are willing to pay $299 for the sort of workup I had.

We're all drawn to novelty, to the promise that there are breakthroughs that will enable us to move forward in huge leaps and bounds. But that's not really where we are when it comes to nutrition—you only have to look at the cycle of dietary fads to know that we humans are still searching for the optimal way to fuel our efforts. That means that in some ways, it's execution of what you know to be true that's most helpful, and not getting lost chasing some magic bullet.

What does that mean? It means handling the basics of nutrition properly every day, instead of eating poorly and then hoping that supplementation can get you over the hump. It means getting good information about your own body and learning how it reacts to what you eat and drink, and even thinking about getting blood work and consulting with a sports nutritionist if you're serious about hacking your diet.

At the elite level, you'd expect that most athletes get these things right. And usually they do. But a surprising amount of time, they don't. Every Olympics you hear stories about athletes who binge on the food—especially the free McDonald's—at the Olympic Village and then have trouble competing. The bewildering array of conflicting advice and choices makes it hard to cut through the noise.

Peter Vint, at the U.S. Olympic Committee, refers to Atul Gawande's terrific book *Better*, which examines the quest to improve medical care. Gawande also remarks on the seductive quality of new research, but notes that "we have not effectively used the abilities science has already given us. And we have not made remotely adequate efforts to change that." For Vint, it's a reminder to ensure that we are diligent in implementing everything we know before we move on to the new simply because of the novelty.

Phil Wagner and his team at Sparta Performance Science in California have seen this with their athletes. Rather than giving them a laundry list of dietary guidelines, they've boiled down their advice for pro athletes to just two things: Make sure that you eat one gram of protein daily for every pound of body weight to support muscular growth, and eat at least eight fist-sized servings of vegetables a day. "You need to get rid of all the other advice to make sure you do the most important things," says Wagner.

That's a lesson all of us—from Wagner's all-star baseball players to the couch potato—can chew on.

4

THE SEARCH FOR EXCELLENCE

Talent Identification, Relative Age, and Why Trauma Might Be a Good Thing

"**D**o you have what it takes to be an Olympic champion?"

That was the question at the top of the flier sent out by the Australian Institute of Sport announcing a new initiative to find athletes who have world-class potential. The AIS has run this sort of program for years, trying to find great athletes early in their careers so the institute can help them develop into champions. The fact that the AIS was looking for great raw athletic material was not surprising. What *was* surprising were the sports involved. In a country that's traditionally been a world power in sports like swimming and rugby, AIS scientists were looking for athletes in judo and boxing.

AIS was born out of failure: At the 1976 Summer Olympics in Montreal, Australians won just one silver and four bronze medals. Worse yet, its tiny neighbor New Zealand won two gold medals, along with one silver and one bronze. When Australia's prime minister at the time toured the Olympic Village, he was booed by the athletes, who felt they hadn't been given the necessary support.

In the U.S. this might have been greeted with a shrug. In Australia, it was a national scandal. Sports, especially international competition, had long been an important component of Australians' self-image. As the country grew from its roots as a British penal colony, its new native-born population used sports to carve out an identity. "Sport in general, and Olympic sport in particular, is one of our few chances to shine on an international stage," said Australian sports historian David Nadel in a newspaper interview. The AIS was formed to ensure that a failure like 1976 didn't happen again. For years, Australia used the research and science that flowed from the institute to become a very successful Olympic nation.

But then there was another disaster at the 2012 London Olympics. After averaging about fifteen gold medals in each of the previous three Olympics, the Aussies only won seven golds in London, while the country's overall medal count fell from forty-six just four years earlier to thirty-five. The Australian misery over those results was only made worse by the triumphant showing by their British hosts—Australia's great sporting rivals won twenty-nine golds.

One thing that jumped out from the results in London was that, of the top ten countries in the medal count, Australia was the only one without a single medal in combat sports. These sports are especially lucrative in terms of medals—because there are so many weight classes, there are fifty-three gold medals available.

You can think of sports as an economics problem—the competition for Olympic gold is a race to maximize resources, from both a human and a financial perspective. For instance, it's been estimated that in Australia, the cost of a single Olympic gold medal is about $37 million. Particularly in a small country like Australia, which has about 23 million residents, making sure that you maximize that investment is crucial. When it comes to bang for the buck, combat sports seemed like a good place to invest, especially since the country would be starting from zero. There was nowhere to go but up.

So the call went out. At eight camps around Australia, prospective fighters came to show what they could do. First, the athletes were measured and then weighed to see which of the seven weight classes in judo or the ten in boxing they would fall into. Then, they performed a vertical-leap test, seeing how high they could reach from a standing jump. After that, there was a twenty-meter timed sprint, to measure lower-body power. Then they progressed to some more specific tests designed by the AIS team to measure certain abilities that could be helpful in combat sports.

Athletes were asked to hop on one foot in a cross pattern: forward and back, left and right. This tested balance. Then, it was on to the upper body: Sitting on the floor with their backs against the wall, athletes tossed a medicine ball as far as possible, and then did as many push-ups as they could in one minute. Finally, there was the dreaded "beep test," a common exercise physiology test that has athletes run back and forth on a twenty-meter course, going faster and faster until they finally can't finish the segment. The beep test helps measure VO_2 max and lactate threshold (even fighters need great endurance).

These tests were a small first step for some of the athletes toward a possible Olympic berth. Who knows, maybe one of them will win a gold medal, and look back at the hopping test as the moment that they demonstrated some part of the physical skills they'd use to get there.

These sorts of efforts are known in the sports world as talent identification programs. In many countries, and in many sports, talent identification is still a random process, based on a combination of a little attention and luck. A gym teacher notices a kid who seems especially fast and encourages him to run track, or a parent notices that her daughter loves to run and jump and signs her up for basketball. Two decades later, they might end up at the top of the world.

But more formal evaluation of athletic potential—especially at younger ages—has become a regular facet of international sports. Throughout the Cold War, the Soviets and other Warsaw Pact countries saw sporting events as an opportunity to demonstrate the superiority of the Communist way of life. Nearly every child was tested at an early age, and those who showed particular promise were shipped to sporting academies, where they trained year-round. (Unfortunately, the athletes were often given performance-enhancing drugs, sometimes without their knowledge.) These programs were funded and closely monitored by the central government.

But talent identification is a balancing act. There are physical and genetic factors that figure into athletic success, as we've talked about. But there are other factors, from the psychological to the sociological. A camp like the AIS combat sports camp can tell you something about the physical ability of an athlete, but it's harder to know which athletes have a high level of internal motivation, or who can handle the stress of competition, not to mention the need to be OK with being punched in the face. Talent ID programs are a blunt instrument, but as more and more countries and sports work to get better at predicting future greatness, we're learning more and more about the types of people who seem to succeed at the highest level, and the circumstances under which they do so.

The Benefits of Being a Late Bloomer

We tend to think of athletic success as a fairly linear process. A kid starts off playing baseball—first T-ball, then Little League. He gets older, plays for his high school team, and if he's good and committed to the sport, maybe he joins a traveling team that roams the country playing other top teams.

If the ballplayer is really, really good, perhaps he's recruited to go play in college or even drafted by a major league franchise. He can then start the climb through the minor league baseball system, and maybe, just maybe, end up in the Show. Maybe he's a marginal big leaguer, or maybe he's one of the top talents, the sort of superstar who will make hundreds of millions of dollars, and even end up with a bust at the Hall of Fame in Cooperstown years later. It's a hike up a mountain, one step after another, until you stand at the very pinnacle.

There's only one problem with this scenario. We're increasingly learning that it just isn't true for most athletes, even the very best ones. For instance, you might think that the performance of a young athlete is the best predictor of his performance when he's older. It stands to reason that if someone can compete internationally as a junior (under the age of eighteen), he's got a big head start over other athletes, and will continue to shine as a senior competitor. Right?

Well, not necessarily. One big factor that contributes to success for younger athletes is faster maturation than their peers. In a study presented at the 2013 American College of Sports Medicine meeting, researchers looked at the careers of athletes who competed in both the finals of the 2000 World Junior Track and Field Championships and in the finals at the 2000 Olympic Games.

Of the athletes who competed in the Junior Worlds, 23 percent went on to win a medal in the Olympics, which is a lot. Obviously, being very good at a young age can be a predictor of future success. But it's more interesting to think of the 70 percent of athletes who *didn't* make it to a Junior World final and went on to compete—and win—at the Olympics anyway. Nearly three-quarters of the best athletes in the world at the senior level weren't among the best in the world when they were younger.

German scientists performed a similar analysis looking at the sport of triathlon. They studied the top twenty in the Olympic Games and World Championships from 2004 to 2011 to see how many had finished in the top twenty of the Junior World Championships in their careers. The numbers were striking: Almost half the men and more than 60 percent of the women who went on to become senior elite athletes didn't have the same success when they were younger. As the researchers flatly stated, "There is no association between rankings in Junior World Championships and on senior elite level."

Why does this matter? Because very often decisions to support athletes, to give them access to equipment and coaches and the resources to allow them to train, are based on those junior results. So if you have an athletic funding system that uses international success as a young athlete as a checkpoint in his or her development—like Germany does—you're likely missing at least half of the athletes in your country who could go on to success when they're older.

There's no art or science to identifying who the best athlete is today; you just have to look at the scoreboard or the results of the race. The trick is to figure out which one will be the best athlete tomorrow—who has the highest upside, the most potential for greatness—and then direct your resources appropriately.

The factors that lead to someone being a standout performer when he or she is ten or fifteen are very different from the skills that lead to elite performance at twenty-five or thirty. Have you ever watched the Little League World Series? In that competition for kids between the ages of eleven and thirteen, it seems like every year, there's a kid who towers over his peers. He'll be half a foot taller than some of his teammates, weigh 50 pounds more, and he's likely the star pitcher and slugger. It's an advantage in maturation, not necessarily in athletic ability. In the junior ranks, it's very often the case that the biggest, fastest, strongest athletes simply dominate, and they do so because they've matured faster. What's interesting is that there's some suggestion that those early-maturing athletes might find themselves at a long-term disadvantage. As Robert Chapman, one of the authors of the study that looked at the track and field athletes, said in a release, "Elite performers in senior sports tend to be the ones who mature later."

When and Where

In many sports, there are age cutoffs that separate young athletes into different years—for instance, in hockey, the cutoff is generally January 1. That means that a kid born on New Year's Day could be a full year older—and more physically mature—than a peer born on December 31, but the two players would belong to the same age group in youth hockey.

This difference in physical maturity is one reason for what's known as the relative age effect, which refers to the competitive advantage that those older players seem to have. In study after study, researchers have found that kids born in the first half of the year have more success than those born in the second half, across sports including soccer, baseball, cricket, and tennis.

The most prominent of these studies have been done on ice hockey—

Malcolm Gladwell built the first chapter of his book *Outliers* around the work of Roger Barnsley, a Canadian psychologist whose early research established just how strong the relative age effect can be. In his studies, he found that roughly four times as many top junior hockey players were born in the first quarter of the year, from January to March, than in the last quarter of the year.

Gladwell uses this information to argue that relative age effect is a huge factor in the possible success of an elite athlete—in fact, of success in any area of life. But there's just one problem. These effects gradually recede, and then seem to disappear altogether, as athletes get older and the developmental gaps between them close.

And there's a further twist. You might be more likely to make it to the pros if you're relatively older. But there are studies that suggest that you might be more likely to become a superstar—one of the very best players in your sport—if you are relatively *younger*.

In a 2007 study, Joseph Baker and Jane Logan looked at the relative age of players who were drafted into the National Hockey League between 2000 and 2005. As the relative age effect would predict, more players were drafted who had birth dates early in the year than would be expected if there were no age effect.

But surprisingly, there was a strong negative correlation between birth date and the position a player was taken in the draft—that is, "relatively younger athletes are more likely to be chosen in the earlier rounds of the draft," a strong indication that NHL teams thought those athletes had more long-term potential.

That finding was echoed by a study of the salaries of players in Germany's top soccer league. Again, there was strong evidence of a relative age effect in terms of the total number of players. When the players' salaries were mapped to their relative age, it turned out that there was a another small, but real effect: The relatively younger players were better paid than the relatively older players. If you believe (as any good economist would argue) that salary is a fair stand-in for ability, it seems that perhaps the relatively younger players were better than the older ones.

There are a couple of hypotheses as to why younger players might excel compared with older peers. To make it to the very top of their sport, younger athletes have shown excellence in a system that appears to be biased against them, by overcoming the disadvantage of their relative age. It stands to reason that they must possess above-average talent to overcome those obstacles.

Then there's the fact that their relatively young age means that they're constantly being physically and mentally challenged by older competitors. There's lots of evidence that these sorts of trials and the frustrations that come with competing against older athletes can lead to younger players giving up on sports. But if you're able to persevere, there's a huge value in being pushed to keep up, whether it's with older peers or even older siblings. The biggest gains in practice seem to come when you're constantly reaching just beyond your current ability.

"I see this all the time," says John Kessel, the director of sport development at USA Volleyball. "The more you play against bigger kids, older kids, even adults, the better you become as an athlete. A disproportionate number of our best players are younger siblings."

So it seems that relative age has a real impact on athletes. But there's another factor that also has a big impact on a young athlete: the kind of place he grew up in. It might be less about when he was born, and more about *where*.

In a study of more than twenty-two hundred athletes in professional hockey, baseball, basketball, and golf, researchers found that athletes who grew up in large cities were much less likely to reach the upper echelon than one would expect from the population distribution:

> The U.S. data indicate that children born in communities of 500,000 or more are significantly disadvantaged in terms of their likelihood of becoming an elite athlete compared with children from communities of less than 500,000. While nearly 52% of the U.S. population reside in cities with populations over 500,000, such cities produce approximately 13% of the

players in the NHL, 29% of the players in the NBA, 15% of the players in MLB, and 13% of players in the PGA.

Other data in the study suggested that very small towns, those with populations under a thousand, were also underrepresented in the ranks of professional athletes. The authors speculate that you need a big enough town to allow for sports facilities and competition; the smallest ones just don't have the infrastructure for young athletes. But once a city gets too large, they argue, youth league and school sports become overly structured.

These geographic factors were much more important statistically than the relative age factors found in the same group of athletes. It's another example of how opportunity is one of the critical elements in developing athletic greatness.

From Trauma to the Top

There's a long tradition of viewing elite athletics as a route out of poverty, a way to use physical skills as a means of breaking away from a situation into which an athlete is born. This stereotyped view is perhaps most prevalent when it comes to the NBA. A few minutes of web searching turns up dozens of profiles of NBA players who "escaped" or "overcame" poverty to make it to the league. One reason for this particular stereotype is the prevalence of African-American players in the NBA. Approximately 75 percent of NBA players are black, compared with about 20 percent non-Hispanic white. (In 2012, 37 percent of black children were living in poverty, compared with 12 percent of white children.)

LeBron James was born to a sixteen-year-old single mother in Akron, Ohio, and they struggled mightily, moving from house to house, couch surfing as his mother bounced from job to job and onto welfare. Eventually, James went to live with a youth football coach in Akron and started to play football and basketball. He began attending school regularly, and laid the groundwork for becoming the global superstar he is today. While it's sadly true that more black children grow up in poverty than do white children, it's simply wrong to infer from these statistics and examples like James that most—or even many—black NBA players come from an impoverished background.

Sociologists Joshua Dubrow and Jimi Adams examined the racial, economic, and familial backgrounds of 155 NBA players born after 1977. They found that the majority of players—regardless of race or ethnicity—came from middle- or upper-income backgrounds: "Looking at the combined categories of upper and middle class of origin reveals that 66 percent of African Americans and 93 percent of whites have advantaged social background." In fact, kids from impoverished backgrounds are underrepresented in the NBA. Despite cases like James's, on balance NBA players come from more affluent backgrounds than the population as a whole.

There was one substantial difference in Dubrow and Adams's study when it came to the familial background of NBA players. Most white players (81 percent) came from a home with two parents, while only 43 percent of African-American players came from a two-parent home. Most African-American players grew up in a home with only one parent, the vast majority of whom were mothers rather than fathers.

Could this be a factor in their later athletic success? In a presentation at the American College of Sports Medicine meeting in 2012, researcher Alan St. Clair Gibson explored the role that broken homes might play in athletic achievement. Rather than viewing a hard childhood as something that an athlete escapes through sports, he suggested that the often resultant anger and resentment is essential fuel for many athletes.

St. Clair Gibson started with a look at three of the best cyclists who have ever raced a bike: Mark Cavendish, Bradley Wiggins, and Lance Armstrong. Cavendish has won more than two dozen Tour de France stages, a world road race championship, two world titles as a track cyclist, and many other top races around the world. His parents separated in Cavendish's early teens, and then his brother was sentenced to prison for a drug charge. Throughout his career, Cavendish has struggled to keep his anger in check, flipping off fans and getting into shouting matches with other riders and teammates. He once told the *Telegraph*, "You have to be a masochist to succeed in the Tour de France."

Bradley Wiggins has won the Tour de France and seven Olympic medals in cycling, and he had a similarly difficult childhood. His parents separated when he was two, and he was estranged from his father (professional cyclist Gary Wiggins). In 2008, his father was found unconscious on the street in Australia, having apparently been beaten at a party, from which injuries he later died. Bradley Wiggins has written about his father's drinking, and also about his own struggles with alcohol and depression when he wasn't in training.

Armstrong never knew his father; he'd call him a "DNA donor" later in life. He was adopted by his stepfather, but was relieved when his mother left him. Throughout his career, Armstrong clashed with other riders and his teammates, and now, of course, we know that he was using performance-enhancing drugs during his seven Tour de France victories.

Three of the best cyclists of the past generation, all from broken homes. All estranged from members of their families. All showing behavior that has ranged from inappropriate to dangerous. Sure, this is anecdotal evidence, but there's something compelling about it. Do difficult life experiences help athletes excel?

Dave Collins, a professor at the University of Central Lancashire and the former performance director for UK Athletics, thinks so. "There is a disproportionately high incidence of early trauma, or at least incidents with the potential to traumatize, in the life histories of elites," Collins and his colleague Áine MacNamara write in a 2012 paper. "The knowledge and skills the athletes accrued from 'life' traumas and their ability to carry over what they learned in that context to novel situations certainly appears to affect their subsequent development and performance in sport." They go on to suggest that talent development programs need to be structured in such a way that they present athletes with challenges along their path through the sport, rather than coddling them. "Talent needs trauma," as they put it.

St. Clair Gibson, in his presentation, wondered just how far would and could we go to find the next generation of talented athletes who have dealt with this kind of trauma. "Would you go and look at defendants?" he asked. While he wasn't quite willing to commit to that, he did suggest some provocative ways to look at making decisions about athletic potential. If you had two athletes who were generally at the same level, St. Clair Gibson suggested that perhaps you'd apply scarce resources and funds to the one who came from the more difficult background.

It's a strange notion, but one that might have some use. There's a fundamental underlying truth to elite athletics that we're often unwilling to comprehend: It's not normal to do what's required to have success at the top of the sporting world. From the years of training to all the things you have to forgo to stay healthy and fit to the simple amount of pain and suffering you must be willing to tolerate, it's not like most people's lives. As Collins and St. Clair Gibson suggest, the trauma that some athletes suffer as children might provide them with the motivation to suffer through these things, as well as give them a certain kind of mental toughness that can be transferred to the sporting realm. Jack Raglin, a researcher at Indiana University, puts it this way: "Elite athletes are very fit, obsessive-compulsive sociopaths."

That's probably not how a proud parent I met in Canberra would view the situation. Louise was staying at the same hotel I was, and we split a ride to the AIS campus, which looks a little bit like a college populated by a distressingly fit student body. Louise was there from Brisbane, along with her thirteen-year-old daughter, who was participating in a national short-course swim meet.

After the meet, her daughter was going to attend a high-altitude training camp, where she would get to work with AIS and national-level coaches, as well as build up her aerobic capacity. "It's amazing the things that they get to do at her age," said Louise proudly. Another important part of the camp, however, is getting athletes on their own and seeing how they handle the situation.

"I've always taken care of her, gotten her gear sorted, made sure that she ate and drank right," said Louise. "But now, she's learning to do that herself. That's what you have to learn how to do to compete internationally." In this mother's discussion of her daughter's growth as a person, Louise touched on an important facet of athletic success.

Marije Elferink-Gemser is a professor at the University of Groningen, in the Netherlands. She and her colleagues there have spent the past decade studying a group of more than one thousand athletes in a variety of sports from soccer and field hockey to tennis and speed skating. This ongoing project, which has become known as the Groningen talent studies, is an attempt to follow the development of these athletes as it's happening, rather than trying to tease out athletic development retrospectively.

The Groningen research has shown that each athlete seems to have a very different developmental curve—that the linear process we imagine is much more fluid. But researchers have also found some commonalities in successful competitors:

> They are known to take responsibility for the progress they make and score higher on aspects of self-regulation of learning, such as reflection and effort. This means that they may set goals that are more realistic and more clear, be more aware of their strong and weak points and be more willing to put effort into training and competition.

When Louise talks about her daughter needing to learn to take care of herself, to take responsibility for her equipment and nutrition, those are important aspects of the self-regulation that Elferink-Gemser writes about. We often think of overbearing parents and coaches pushing an athlete to excel in a sport, but the truth is that it's very rare to find a great sportsperson who doesn't find the bulk of his or her motivation from within.

Finding the Right Event

Mo Farah and Usain Bolt were the stars of the London Olympics, with Farah winning the gold on the track at 5,000 and 10,000 meters, while Bolt swept the 100 and 200 meters. Beyond their success, it was their charisma that electrified the crowds at the stadium and around the world: Farah with his arms over his head in the shape of a heart in his "Mobot" celebration, and Bolt with his arms extended like an archer in his "To Di World" gesture. Late in the games, they even celebrated together, each doing the other's signature move to a delighted roar from the fans.

They were linked in London, but they're at opposite ends of the physical and physiological spectrum. Farah is 5 feet 5 inches tall and weighs 128 pounds; Bolt is a full foot taller at 6 feet 5 inches, and weighs 190. Farah is an aerobic machine; Bolt is a scorching speed demon.

That's why the challenge that Farah issued to Bolt in the summer of 2013 was so much fun—he suggested that the two stage an exhibition race over an intermediate distance as a fund-raiser for charity. This immediately brought up a fascinating question: What would be the best distance for these two to run to ensure the most competitive race possible?

Bolt has spent his athletic career developing his muscles to contract very rapidly for a very short duration. It's not about the ability to sustain the effort or the energy the muscles need to fuel it; it's about putting out as much energy as possible as quickly as possible. That sort of effort is fueled by the anaerobic energy system, which can produce a burst of energy very quickly, but not for long.

Farah's just the opposite. He relies primarily on aerobic energy in his races. At that lower intensity level, you can maintain your performance for hours, since there isn't the same buildup of waste products. But those aerobic pathways don't produce enough energy to fuel the super-high-intensity efforts in which sprinters specialize.

To clarify, both distance runners and sprinters rely on a balance of aerobic and anaerobic energy during their events—it's never 100 percent one or the other. But that balance changes as the events change. Short events are predominantly anaerobic; long events are mostly aerobic. Somewhere in the middle, there must be a transitional point. But where?

Australian researcher Rob Duffield has published a series of papers that try to clarify the relative contributions of each energy system in races ranging from 100 meters to 3,000 meters. Duffield has found that at 100 meters, energy is 21 percent aerobic and 79 percent anaerobic. For the 3,000 meters, that's roughly flipped, with 86 percent aerobic and 14 percent anaerobic. If you take Duffield's work and graph the various points, you start to get an answer as to where the shift from anaerobic to aerobic energy occurs: right around 600 or 800 meters. Which just happened to be the distance of the race to which Farah challenged Bolt.

The race hasn't come to fruition yet, but the mental exercise of trying to predict what would happen is familiar to people working in the talent identification world. From a limited amount of information, they have to make (highly educated) guesses about the outcome of an event and an athlete's adaptation to it. Traditionally, this type of evaluation has been focused on athletes before they reach the elite level, but increasingly, teams and countries are coming to realize that there's another way to approach the problem.

The traditional approach to talent identification looks at a teenage runner's body type and physiology and uses that to try and determine what type of running she would be most successful at, then funneling her into either sprinting or distance running. It's an outlook that tries to find and groom athletes for international success based on their prowess at the sport they're already competing in.

A more efficient way to garner gold medals might be to look beyond where an athlete is today—not just in terms of his physical and mental development and trying to project that into the future, but in terms of the sport he's competing in. You might have lots of potential champions available to you, if only they were involved in the sport that best suited them.

5

THE FAST TRACK TO GREATNESS

Talent Transfer and the 10,000 Hours Rule

As Helen Glover and Heather Stanning settled into their rowing shell for the finals of the women's pairs, the pressure on the two athletes reached a previously unimaginable level. Over the course of the first four days of the London Olympics, no athletes from Team Great Britain had won a gold medal, and the country was in the throes of a tabloid-fueled panic attack. Several heavy favorites in other sports had seemingly collapsed under the pressure of competing in front of their home fans, and the tone of the coverage had taken on a neurotic cast. When, the country seemed to ask itself, will we finally win gold?

It was a windy, overcast day at the Olympic rowing venue at Eton Dorney, a man-made rowing lake about twenty-five miles west of London. Glover, a former gym teacher, and Stanning, a captain in the British Royal Artillery, tore away from the start of the race, setting a furious pace. By the five-hundred-meter mark—a quarter of the way through the race—they were more than a boat length ahead of the rowers from New Zealand, who had beaten Glover and Stanning at the 2011 World Championships. After that defeat, the Brits had focused on increasing their stroke rate—taking more strokes in a given period of time, which, if the pace can be maintained, would lead to a big increase in speed.

Rowing is a sport that requires an immense amount of strength and cardiovascular fitness, but also a level of grace and teamwork. In a pairs boat, the two rowers have to find a rhythm that allows them to work almost as if they are one body, keeping the shell headed straight down the course. When it's going well, there almost seems to be a lightness to it, even as the athletes' chests heave as they try to get oxygen to their lungs, and their arms and legs scream against the strain.

That's the state that Glover and Stanning entered. Their boat looked like it was somehow skimming over the water as their competition bogged down. Thirty thousand fans screamed and screamed as the pair passed each time check with a larger lead. As they entered the last 250 meters of the race, something funny happened. Helen Glover started to break into a smile. Watching a replay of the race, you can see her look from side to side, a grin coming over her face even as she kept driving toward the finish line.

It's the sort of smile you could interpret as the relief that an athlete might feel as a lifetime of hard work is validated by an Olympic gold medal. And certainly, Helen Glover had worked amazingly hard as a rower to capture gold.

But the lifetime part? Not so much. In fact, only four years earlier, Glover was falling out of the boat on her first attempts to maneuver one on the water. Rather than the culmination of a life's work by Glover, her gold medal and the nation's joy over it was—in part—due to some very clever ideas from British sport scientists and the power of newspaper advertising. Helen Glover wouldn't have been in that boat at all if not for an ad her mother saw in a local newspaper. Team GB was looking to capture sporting glory, and they were running a casting call for athletes. The only requirement? That you were tall.

You Must Be This Tall to Win the Race

Walking around the Olympic Village is a visual catalog of the extremes of the human race. Once you put athletes from all over the world and dozens of sports together, you see that athletic greatness comes in every size, shape, and color imaginable. The juxtaposition can be almost comical. One moment, you see nearly pocket-size female gymnasts, young women who might be just 4 feet 11 inches and 90 pounds, like Olympic champion Gabrielle Douglas, but who have an immense amount of strength and flexibility. The next moment, you see a comically gigantic volleyball player, like Russia's Ekaterina Gamova, a towering two-time silver medalist who stands 6 feet 8 inches and weighs 180 pounds. It's a testament to the diversity of both the human race and the games we play that these two women, who barely look like they're part of the same species, can both reach the top level of athletic competition.

The massive morphological diversity in elite athletes is a relatively recent development. At the turn of the twentieth century, there was a more universal idea of what an athlete looked like, an almost Platonic ideal of a well-muscled (but not too big) man with good endurance (but not too much, since that would require a reduction in body mass). Across many sports, athletes tended to cluster around the same size. A historical study of world-class male athletes in twenty-one sports demonstrates the changes in the past century. In 1925, athletes in fifteen of those sports all clustered between 5 feet 7 inches and 5 feet 11 inches, and 132 and 176 pounds. Today, only eight sports (mostly running, cycling, and gymnastics) have elite athletes who fall in that zone. In the history of the Olympics, forty-two athletes have won medals in more than one sport, according to the IOC's research department. But only nine of those have happened since 1936. As sports scientist Timothy Olds writes about the Olympics, "Such improbable cross-sport successes as diving and water polo (American Frank Kehoe in 1904), boxing and bobsled (American Eagen Edward 1920–1932) or water polo and fencing (Belgian Boin Victor in 1908–1920) have become morphologically impossible."

There are certain sports in which a specific body type is almost a prerequisite for success. Why are gymnasts like Douglas, divers, and figure skaters almost always shorter than the average population? Because those sports require rapid rotations of the body to execute twists and somersaults. Those athletes need what a physicist would call "high angular acceleration." The shorter the thing is that you're trying to spin, the faster it will rotate given the same force. Smaller athletes can spin faster, and that's a big competitive advantage.

On the other end of the spectrum are volleyball players like Gamova. The net in women's volleyball is 7 feet 4 inches. If you're not tall enough to get your arms well above the net when you're spiking or trying to block the ball, you're not going to be effective. Gamova, who can reach well over ten feet high when she jumps, has little trouble clearing the net. Being tall can also be an advantage in a sport like tennis. Height is most advantageous during the serve, in which the geometry of the game means that each additional centimeter of height gives a player an extra four centimeters of the opponent's court to attack with a serve.

Basketball is the most obvious example of a sport in which extreme height can be a central factor for success. While it's true that players as short as 5 feet 3 inches tall have played in the NBA, they are remarkably rare. On the other hand, being exceptionally tall can sometimes seem like an invitation into the league. In 2011, *Sports Illustrated* estimated that there are likely fewer than 70 men in the United States between 20 and 40 years old who stand 7 feet or taller. But of those men, almost 1 in 5 have played professional basketball in the NBA. In comparison, the probability that a very, very tall man (between 6 feet 6 inches and 6 feet 8 inches) would make it to the NBA was just 0.07 percent.

Rowing is another sport in which being tall is a significant help. The physics of rowing a boat means that having what biomechanists refer to as a "long lever" allows an athlete to apply force to the oar for a greater period of time during each stroke, provided that he also has the muscular ability to apply that force. That's why elite rowers tend to be taller than the general population. But it's also important that they are relatively light for their size, because carrying too much extra weight slows down the boat. In fact, rowing is one of the most morphologically constrained sports on earth, in terms of the ideal athletes being physically rare in the general population. Even NFL players, whom we think of as physical outliers, are more similar in height and weight to the general population than elite rowers. (It's the combination of being tall but light that's so uncommon.)

Tall is good. And tall is rare. Tall and athletic is rarer still. That's why UK Sport launched the delightfully named Sporting Giants program in February 2007. The pitch from the sports scientists behind the program was simple: Are you a man who's taller than 6 feet 3 inches or a woman who's taller than 5 feet 11 inches? Do you have an athletic background? If so, UK Sport wanted to hear from you to see if you had what it would take to be a rower, a volleyball player, or a team handball player—another sport in which height matters.

That's the advertisement that Helen Glover's mother saw in the newspaper. Glover had been an accomplished athlete at the county level, and had even played field hockey on a second-tier English national team, but she had never made it to the full international level. Standing 5 feet 11 inches, she also met the height requirement. She submitted an application to the program—along with more than four thousand other athletes.

About fifteen hundred of those athletes were then selected to undergo physiological and skill-based testing. Glover recounts the testing experience on the UK Sport website: "I remember sitting in a room in Bisham Abbey [the testing center] and someone saying: 'A gold medalist in 2012 could be sat in this room. Look around you.' I thought: 'Right, I'm going to make that me.' It was quite surreal."

Sixty-nine women and men were brought into an eight-week program for rowers, in which they trained full-time with national team coaches and scientists, undergoing constant evaluation of their training responses. At the end of that phase, cuts were made and the smaller group continued on their way.

The results were spectacular. In the first year, seventeen of the athletes identified through the program reached national finals in their events. And by the time the London games rolled around, eleven of them competed for Team GB, lead by Glover, who never would have sat in a racing shell without the program, let alone ended up standing at the top of the Olympic podium.

No Magic Paths

"But wait a minute," I can almost hear some of you saying. "I'm pretty sure that you can't reach an elite level in anything without lots and lots of practice. Haven't you ever heard of the 10,000-hour rule?"

The 10,000-hour rule, as it's commonly called, stems from the work of Anders Ericsson, a psychologist who has spent most of his career trying to explain the factors that lead to expert performance. His most famous paper, "The Role of Deliberate Practice in the Acquisition of Expert Performance," was published in 1993, and has been cited more than four thousand times in the academic literature since. Ericsson boiled down the essence of his forty-four-page paper to a single sentence in the summary: "Many characteristics once believed to reflect innate talent are actually the result of intense practice extended for a minimum of 10 years."

Ericsson's paper centers around two studies of musicians. In the most famous, several groups of violin players were assembled at the Music Academy of West Berlin, sorted by their perceived level of potential as rated by the teachers at the academy. The researchers attempted to reconstruct the musical histories of the violinists by interviewing all of them on topics ranging from when they took up the instrument to how many instructors they had studied with to how often they practiced.

What they found was that the better violinists practiced more than the less accomplished players. Much more. "The central claim of our framework is that the level of performance an individual attains is directly related to the amount of deliberate practice," Ericsson writes. That is to say, the more you practice, the better you will get at something, and it's very improbable to be great without that practice.

Ericsson's theory exploded into mainstream consciousness when Malcolm Gladwell wrote about it in *Outliers*. After discussing Ericsson's studies, Gladwell writes:

The idea that excellence at performing a complex task requires a critical minimum level of practice surfaces again and again in studies of expertise. In fact, researchers have settled on what they believe is the magic number for true expertise: ten thousand hours.

This belief has echoed through popular science writing, with books such as Geoff Colvin's *Talent Is Overrated* and Matthew Syed's *Bounce* continuing to popularize the belief that ten thousand hours is a magic path to greatness in any field. But examine that claim in the context of elite sports, and it just doesn't hold up. Helen Glover went from rowing novice to Olympic gold medalist in four years—far less than the ten thousand hours (or ten years of full-time practice) that Gladwell and Ericsson would predict as necessary.

Then there's Donald Thomas. He was a basketball player at tiny Lindenwood University, in Missouri, known for his dynamic leaping ability on the court. One day, some of the guys on the school's track team and Thomas were ribbing each other, and finally a bet was made that Thomas couldn't high-jump 6 feet 7 inches. Off the group went to the high jump pit, Thomas still wearing his basketball shoes and never having high-jumped before.

He cleared 6 feet 4 inches. Then 6 feet 7 inches. Then 6 feet 8 inches. Finally, the bar was set to 7 feet, and he cleared that too.

"That was Thursday, Jan. 19, 2006," Thomas told Brett Hess at CSTV .com. "We ran over to the track coach's office. On Saturday, I jumped 7 feet 3¼ inches to win a meet at Eastern Illinois."

Thomas graduated from Lindenwood and then went to Auburn University for grad school, where he could compete at the highest level of collegiate track. In March 2007, a mere fourteen months after his first high jump, he won the NCAA indoor national high jump title. Five months later, less than two years after that day he was joking with his friends on the track team, Thomas won the World Championships in Osaka, Japan, jumping 7 feet 8½ inches to defeat a field filled with competitors who had been jumping for more than a decade.

Now, Thomas is, to borrow a term from Gladwell, an outlier. It's not as if there are hundreds of examples of athletes going from being a complete novice at a sport to a world champion in two years. But as you dig into the athletic history of international-class athletes, you start to find that it's not all that rare to reach the top level of competition with far less than ten years of experience in your chosen sport.

Researchers have looked at sports ranging from wrestling to field hockey to soccer and found that the usual training requirements to compete at the international level tend to be much less than 10,000 hours. For the wrestlers, it was about 6,000 hours, while the field hockey players took about 4,000 hours, and the soccer players around 5,000. Interestingly, these amounts of practice accumulated over the same ten-year time frame that Ericsson proposes, but there are far fewer active hours involved, as compared to the musicians in Ericsson's original study.

A 2004 study looked at the developmental history of 459 Australian athletes who represented the country in international competition. The researchers found that the average number of years it took from when an athlete started a sport to when he or she made it to the senior national team was seven and a half years. The numbers are even more interesting when you drill down a little more. While the average was seven and a half years, there was a lot of variation around that number. One large group of athletes (28 percent of the total) developed very quickly, going from novice to elite in less than four years. On the flip side of that coin was an equally large group (30 percent) who took more than ten years from starting the sport to make a national team.

The important differences between those two groups might be the difference in their sporting histories. The ten-year group began their main sport at a young age (around eight years old), and hardly played any other sports before picking that main activity. They focused on the one sport for a long time and got very good at it. In their case, Ericsson and his devotees are correct—one path to expert performance is a long period of intensive training.

The fast-developing athletes followed a very different path. First, they were more often found in individual sports rather than team sports. This makes some intuitive sense, as an individual sport often allows for a less complicated learning curve. A runner's or rower's performance largely depends on his or her physiology—things like VO_2 max and aerobic thresholds, which we discussed earlier. Certainly there are tactics involved in racing, but they're relatively straightforward. But mastering a sport like soccer at the elite level doesn't just require great physical fitness and skills; it requires a deep understanding of the game and its tactics and strategy. You don't just need to be able to run fast and dribble and kick the ball. You have to understand how to combine your play with that of the other ten players on your team.

In fact, there's a rich vein of research in team sports that focuses on the importance of tactical knowledge. How important are these skills? Dutch researchers studied a group of elite youth players, and used a tactical evaluation to rate a player's ability to correctly position himself and make quick decisions on the field. The players with high scores on this ability were almost seven times more likely to land a professional contract as those who scored nearer the bottom—and remember, these were all top eighteen-year-old players.

But the more important difference between the fast- and slow-developing athletes is how and when they came to their primary sport. The fast developers played many more sports before they started their main sport, which they started much later in life—around seventeen years old. Although they had spent their childhoods trying out various sports, once they found the one that clicked, the great fit between athlete and sport allowed for their very quick progress.

The huge popularity of the 10,000-hour theory has led to some real changes in how parents are managing their kids' athletic careers. Ericsson writes, "Consistent with our hypothesis, we find that the higher the level of attained elite performance, the earlier the age of first exposure as well as the age of starting deliberate practice." (Ericsson's first study might have been about musicians, but he has argued that the same rules apply across any elite performance.) While studies like we've just examined undercut this position, there are still lots of parents who bet that the 10,000-hour theory is the best possible way they can help their kids become great athletes. The popularity of the theory has caused a shift away from the more fluid model of sports participation we saw a generation ago toward a world where kids pick a single sport at a young age and focus on it to the exclusion of others.

It's important to remember the odds that parents are playing here. Let's take baseball in the United States, which is, relatively speaking, an "easy" sport to play at the highest professional level, given that there are thirty Major League Baseball teams, each with forty-man rosters—twelve hundred spots available. In 2012, the National Sporting Goods Association estimated that 5.5 million kids between the ages of seven and seventeen played youth baseball. So, if you really do the math, that means the MLB player population is 0.022 percent the size of the youth baseball population. Or, to think of it another way, an individual kid's odds of making it onto an MLB roster is about 4,600 to 1. Chances of making it to the NBA are even slimmer, given that there are only 440 NBA jobs available. The odds there work out to about 21,000 to 1. And that's not even factoring in the foreign players in all these sports, who make the actual odds much longer.

These long-shot odds haven't deterred lots of people from not only dreaming about their kid making it as a professional athlete but also actively taking steps to try and make it happen, most prominently, by focusing their child on just one sport at an early age. You can see the results of that focus almost everywhere you look in youth sports. The number of ligament repair surgeries done on teenage baseball players has mushroomed as they pitch more and more innings. Overall, more than 3.5 million kids under the age of fourteen are injured annually playing sports. Nearly half of those injuries are caused by overuse, according to statistics from the American Orthopaedic Society for Sports Medicine.

A recent study by doctors at Loyola University suggested that a good rule of thumb is that a young athlete shouldn't spend more hours per week than his or her age playing one sport. So, if you're twelve years old, you should spend twelve or fewer hours playing a particular sport. Athletes who exceeded this guideline were 70 percent more likely to suffer overuse injuries, according to this research, which was presented to the American Medical Society for Sports Medicine by Dr. Neeru Jayanthi.

So if you're looking to help your kid in his or her athletic endeavors, try to resist the lure of popular notions like the 10,000-hour rule and the importance of relative age, which we covered in the previous chapter. Some parents I know have "redshirted" their young boys as they were about to head into kindergarten, so that they will be old for their class. At the same time, they started them on focused baseball training to the exclusion of any other sport. As we've seen, that approach could lead to a great athlete. But the science isn't a home run—the majority of those elite Australian athletes played lots of different sports as a kid, rather than focusing on one. The longer the wait before specializing in one sport, the better chance that a young athlete will find a sport that he's not only best suited for but also enjoys the most. German researchers recently reinforced the importance of this sampling in a study over 1,500 German national team athletes. "A larger proportion of the world class athletes reported to have changed their main sport during their career," the researchers write, "and the proportion of world class performers was highest when the athletes had experienced more sports."

In fairness to Ericsson, some of the ideas that have flowed from his research can't actually be attributed to him. For instance, Ericsson has never used the phrase "10,000-hour rule" in any of his writing. He wrote a paper in 2013 addressing some criticism of his work, in which he says that he has never claimed that "deliberate practice is *sufficient* to explain the acquisition of all aspects of expert performance." (In the same paper, he refers to the "popularized but simplistic view of our work circulated on the internet," which brings to mind a couple of bloggers rather than a stack of best-selling books.)

As Ericsson states, the point of his seminal paper was that "the best group of violinists had spent significantly more hours practicing than the two groups of less accomplished groups of expert violinists, and vastly more time than amateur musicians." In other words, practice was directly related to accomplishment in the musical arena (but that it doesn't prove a causal relationship between the two). But that same direct relationship between practice and accomplishment just doesn't hold in sports, especially when you put a group of scientists to work trying to short-circuit the process.

Hitting Fast-Forward

How do you accelerate the development of world-class athletes? Can you take specific steps to help speed athletes from novice to expert, instead of waiting for years of training and effort to kick in?

Those were the questions that scientists at the Australian Institute of Sport asked themselves when skeleton—a sport that involves hurtling head-first and facedown on a bobsled track on a small sled—was reintroduced to the Winter Olympic Games. Needless to say, Australia didn't have much of a skeleton tradition, or even a single bobsled track in the entire country.

In fact, there were fewer than one hundred registered female skeleton athletes in the world at the time, which the AIS researchers saw as a real opportunity. As they wrote, "Skeleton offered a unique opportunity to identify a complete novice or target existing high-performance athletes to transfer into this sport. The ability to utilize existing high-performance athletes' skills and competition experience gave us the capacity to exploit talent gaps at the World Cup level by possibly compressing the developmental time frame of these athletes."

The first step for the team was to analyze what made athletes successful in the sport. In their analysis, they found that a single crucial variable accounted for up to half of the variance in the final time of a run. It wasn't the skill of the pilot or the design of the sled. The most important thing was the athlete's start time. The faster a competitor pushed the sled through the thirty-meter start zone before jumping on it, the better she performed.

So researchers set up a national testing campaign, looking for women with backgrounds in competitive sports who clocked great times when they were tested on a thirty-meter sprint. They also evaluated candidates to see how well they responded to feedback and coaching. Eventually, they picked a group of ten athletes, including track sprinters, a water skier, and several surf life savers, a popular Australian competition for lifeguards that requires sprinting through sand to unearth flags buried on the beach. These athletes were trained in skeleton, and then the four-person team was selected.

The athletes were given access to the best coaching, equipment, and sports science. Every training and competitive run was analyzed and dissected, as coaches looked for places to improve start and steering techniques. Specialized strength and conditioning programs targeted explosive power and sprint speed. The program was a success: An Australian racer qualified for the Olympics just eighteen months after she first saw a sled. Amazingly, she had completed only 220 runs before doing so. (A typical U.S. skeleton racer makes upwards of two thousand runs before appearing in the Olympics.)

In the paper describing the skeleton program, AIS researchers called their approach "deliberate programming," to contrast it with Ericsson's theory of "deliberate practice." Their differing approaches are detailed in the following chart from their paper:

	Deliberate programming	Deliberate practice
Basis of talent	Innate abilities	Acquired by practice only
Available talent pool	Highly limited	Unlimited
Existing athletes	Can be leap-frogged	Cannot be leap-frogged
Prior sporting experience	Favored	Not favored
Developmental time	Short (2–3 years)	Long (10+ years)
Competition	Essential	Not considered
Specialization	Late	Early
Fast-Tracking	Achievable	Not achievable
Enjoyment	Important	Not important

This talent-transfer model offers a third path to elite performance, which straddles some of the divide between fundamentalists on either the genetics/talent or practice side of the debate. Like many debates—scientific or otherwise—the one between nature and nurture when it comes to elite athletes has become polarized to an extent that serves almost no one's interests in the real world of competition.

It's not surprising that the skeleton study was done at a place like AIS, which is focused on the very specific goal of winning medals at the world level. There's a difference between work that's solely academic in focus and work that's meant to lead to competitive results. While scientists at AIS and its competitors around the world might publish and lead research in their fields, the primary goal is much more pragmatic: to win.

Many Roads Lead to the Top

The practical benefits of talent transfer are what make it such a powerful tool. That's why UK Sport put out the call for tall athletes, which led to Helen Glover's gold medal, as well as a bronze in the World Rowing Championships from her teammate Vicky Thornley. And that's why they're running other programs to take athletes who might have failed to reach the top level in more popular sports like soccer and see if they might thrive in other sports. It's an outlook that an economist would view as a maximization of resources—in this case, ensuring that talented athletes have the opportunity to find their best chance to succeed at the highest level.

Sometimes, that best chance comes by adapting to an athlete's changing circumstances and body. At the Athens Olympics in 2004, Rebecca Romero was part of the silver-medal-winning crew for Great Britain in the women's quadruple sculls, and won a gold in the World Championships the following year. But in January 2006, she had to retire from rowing due to persistent problems with her back, possibly the result of nearly a decade's training in rowing.

Romero didn't want her athletic career to end, so she looked for a sport that would put less stress on her back, and settled on cycling. She went in for testing with British Cycling, and her power numbers were off the charts. "They said I had one of their best results ever," Romero told *Wired UK*. "It was insane. I didn't have the physiology or the bike skills. Very quickly I was being taught everything and I was going from a nothing to being a member of the team and aiming for an Olympic medal."

So Romero set her sights on a rare feat—winning a second Olympic medal in a completely new sport. With just two and a half years before the 2008

Beijing Olympics, Romero and her coaches devised a plan by working backward. They started with the times they assumed would be required to medal in Beijing, and then they plotted out a training program that aimed to deliver those performances through a steady path of improvement. Their assumptions were right on, as was their training plan; Romero won gold in Beijing in the women's individual pursuit track cycling race. "People talk about the ten years it takes to achieve something, but I believe you can accelerate the learning process if you are smart about the way you practice," she said. "It isn't just about the hours and hours of training; it's all the tiny things that add up." (She clearly learned about those marginal gains from Dave Brailsford.) Romero retired from the British cycling team in 2011 after the pursuit was dropped from the Olympic program, but she's hardly taking it easy—she's since taken up triathlon, and qualified for the Ironman World Championship in Hawaii by winning her age group in her first-ever Ironman race.

It's crushingly obvious that Romero has exceptional physiology. But here's where Ericsson's arguments crash headlong into genetic differences between individuals. For instance, he believes that even the sort of endurance physiology that Romero has is exclusively the result of training:

> The activation of genes is critical for developing physiological adaptations of the body and nervous system that enable expert performance in the particular domain. However, so far the scientific evidence of genetic mediation suggests that healthy children seem to have the critical genes required for the desired changes as part of their cells' dormant DNA. A recent review found that individual differences in attained elite performance cannot, at least currently, be explained by differential genetic endowment, barring only a few known exceptions of characteristics that directly mediate the performance, such as body-size and height.

That is to say, Ericsson believes that practice alone can give you the ability to perform at the elite level, and that every healthy child has that ability as part of his or her DNA. But from our look at genetics, we know that's just not so. Remember, research from people like Claude Bouchard shows that about half of the difference in an individual's starting aerobic fitness is determined genetically, and, more important, that there are large differences in individual responses to training.

If you're going to hold on to the belief that only practice can lead to elite performance, you have to find a way to explain stories like Evelyn Stevens's. Stevens worked as an analyst for Lehman Brothers on Wall Street, spending countless hours at the office, making countless dollars. She had been an athlete growing up and played varsity tennis at Dartmouth, but sports had fallen by the wayside. Her days were filled with PowerPoint presentations and finance-related matters.

Then, in the fall of 2007, she went to visit her sister in California, who was a bike rider, as was her sister's boyfriend. They took Stevens for a sixty-mile ride up Marin County's famed Mount Tamalpais, and Stevens somehow kept up. The next day, they went to a bike race in Golden Gate Park, and Stevens competed on a bike for the first time. There were several crashes, but she finished.

Back in New York, Stevens bought a low-end road bike and started riding. She checked out a racing clinic for beginners, and had a blast. Soon, she was entering races as a Category 4 rider (the lowest amateur classification) and winning them.

Then came the Green Mountain Stage Race in Vermont in August 2008, less than a year after Stevens had started riding. The pro racers started five minutes ahead of the group that Stevens was in, but that didn't matter. The woman who had been racing bikes for less than four months closed that gap alone, and finished near the front of the pro race—the cycling equivalent of a recreational jogger starting a 10K five minutes after elite runners and going on to beat most of them. The next year, she won the Tour of the Battenkill, the largest one-day bike race in America, as well as the Cascade Cycling Classic, one of the most prestigious races in the country. By 2010, she was the U.S. National Time Trial champion.

Today, she's one of the top female riders in the world, having won the team time trial in the UCI Road World Championships and competed in the Olympics. Given her results thus far, and her continued improvement as she develops as a cyclist, Stevens is one of the favorites for gold in the 2016 Olympics in Rio. Although practice will continue to help her develop, it's clear that there's something else at work, since she became a world-class cyclist in less than a year after taking up the sport, with something more like a thousand hours of training than ten thousand.

It's not hard to understand why the 10,000-hour model is so seductive, because it tells us something that most of us would like to believe. The model says that, essentially, anyone can do anything if they just work hard enough. That's a nice thing to think. When we lean down and say to our kids, "You can be anything you want to be," we want that to be true, whether it's becoming the President of the United States or an Olympic gold medalist.

To put it in the language of a logician, the most devoted acolytes of the 10,000-hour theory believe that ten thousand hours of deliberate practice is both necessary and sufficient to become an expert performer. That is to say that no one becomes an expert performer without ten thousand hours of work, and that anyone can become an expert in anything with the right amount of the right kind of practice.

Putting aside the science, this doesn't even stand up to casual observation. If the rule is right, then the only thing that matters is the number of hours you've practiced. There would hardly be any reason to have athletic competitions at all—just have each competitor submit a training history, and the person with the most hours spent in deliberate practice could be declared the winner. And if ten thousand hours is necessary and sufficient, then we can't account for all of the cases in which athletes have reached the top of the podium without it.

There's a massive diversity in the paths that athletes take, from the seemingly effortless rise of a prodigy to the career-long struggle of a marginal athlete. What a practice-based model like Ericsson's points out is that effort and opportunity are hugely important when it comes to making a champion. Even the success of that prodigy is the result of a huge amount of work.

A message that hard work is crucial to achievement is great. A message that hard work is often underrated is totally fair. But a message that hard work is the *only* relevant factor in becoming an expert is simply wrong. Biologist Richard Lewontin tries to explain the interaction between genetics and environment by using the metaphor of a bucket: "Genes determine the size of the bucket, and environment determines how much is poured into it."

Sports scientists Ross Tucker and Malcolm Collins, in a review of the relative contributions of genetics and practice to athletic achievement, propose a model in which "training maximises the likelihood of obtaining a performance level with a genetically controlled 'ceiling.'" Some sports might rely more on genetic factors—say, those in which physiology is more dominant—while those in which skill is more prominent might trend toward practice. But every great athlete is the product of the interaction between genetics and effort—the raw materials they were given as athletes, and how much they've developed them.

So where does elite performance come from? Talent or practice?

The answer, of course, is both. The greatest athletes are born, and then made.

6

LEARNING TO BE THE BEST

Bad Pass Fridays, Loving the Camera, and How the Game Teaches the Game

Walking down the row of golfers at the U.S. Open's driving range is a visual lesson in practice and skill refinement. Every player has a unique routine that he follows before a round starts, a specific set of drills and shots that he hits. You see some players quickly work their way through their bags, hitting a couple of shots with each club to get the feel for their swings and their bodies that day. You see elaborate preshot routines and more casual approaches.

But mostly what you'll notice is an almost metronomic consistency that seems to approach perfection. On the range before a round at the 2012 U.S. Open at the Olympic Club in San Francisco, German pro Martin Kaymer took out his pitching wedge and started hitting shots at a flagstick about 150 yards away, working on his draw.

Most PGA pros rarely hit the ball straight. They either hit a fade, in which the ball curves left to right during its flight, or they hit a draw, in which it curves right to left. Almost every golfer has a natural shot shape, one that's organic to his swing, but the best players are able to work the ball both directions, which allows them to deal with doglegs and pin placements that might favor one flight or the other. Part of the reason that Tiger Woods's game in the early 2000s was the most dominant golf ever seen was his ability to hit whatever shot was required—draw or fade, low or high trajectory—in any situation.

Kaymer had won the PGA Championship, one of golf's major titles, in 2010, and early in 2011 he climbed to the top of the world golf rankings. But he decided that his game was missing something—the ability to consistently hit a draw. Kaymer desperately wanted to win the Masters—which is thought by many to be golf's most prestigious tournament—but the course at Augusta National Golf Club greatly favors a player who can draw the ball, especially off the tee. (That's for right-handed players. Lefties who play a fade have the same advantage, like Masters winners Phil Mickelson and Bubba Watson.)

"Every time I left Augusta, I was very frustrated, not because I had just missed the cut but the way I missed the cut, because I had no idea how it feels to hit a draw," said Kaymer at a press conference at the 2012 Ryder Cup. So Kaymer set out to capture that feeling . . . and his game seemed to fall apart. He worked on hitting the draw, but lost the ability to consistently hit his more natural fade. He bombed out of the 2011 Masters, missing the cut, and lost his spot at the top of the world rankings, eventually falling all the way to thirty-third in the world.

I'm a 12 handicap, and several times a round, I'll hit a nearly perfect shot—a drive striped down the middle of the fairway, or an 8-iron that tracks the flagstick the entire way. But of course, I make horrid shots as well, from shanks to hooks to fat shots that barely seem to go anywhere at all. The difference between me and a good golfer is how often we hit those two types of shots. A scratch golfer will hit those good shots much more often than I will, and will have far fewer of the horrible ones. And the difference between that scratch golfer and a PGA Tour pro is another massive gap. The tour player doesn't just consistently hit great shots. He consistently hits great shots under a huge amount of pressure.

It took Kaymer almost two years to get his swing back. Watching him hit those pitching wedge shots on the range at the Olympic Club, it was hard to imagine that he ever struggled at all. Every swing was identical, an effortless-looking turn away from the ball and then back through. Each shot arced off the club face in a high flight, a gentle right-to-left draw tracing through the sky that landed with a soft thump, shockingly close to the flagstick. Kaymer hit thirty balls in a row, and you could have nearly covered the results with a king-size bedsheet—that's how precise and consistent the shots were. He would go on to finish fifteenth in the U.S. Open that week, his best result since making his swing changes. A few months later, he would win the match that clinched the Ryder Cup, giving Europe a stirring comeback victory over the United States in golf's top team competition.

Then, in the 2014 season, Kaymer returned to the pinnacle of the sport. In May, he won the Players Championship, the most prestigious tournament on the PGA Tour outside of the major championships. And in June, Kaymer went to the U.S. Open and completely torched the field at the famed Pinehurst No. 2 course. In a week where only three players finished under par, Kaymer started by shooting two consecutive rounds of 65, and ended the tournament at 9 under par. His 8 stroke margin of victory was the third most dominating performance in the 114-year history of the championship. Afterward, at his press conference, Kaymer reflected on the difficult process of rebuilding his game, and how good it felt to win.

"You want to win Majors in your career, but if you can win another, it means so much more," said Kaymer. "Some people, especially when I went through that low, called me a one-hit wonder. So it's quite satisfying to have two under your belt. And I'm only twenty-nine years old, so I hope I have another few years ahead of me."

Kaymer isn't the only golfer who looked for ways to tweak his game when he was seemingly at its top. No player has done so more than Tiger Woods, who has completely rebuilt his swing three times during his career. But Woods's success in retooling is the exception, not the rule. The history of golf is littered with players who tried to add a shot or change their game, and never reached the same heights as before. David Duval, Seve Ballesteros, Ian Baker-Finch—they are all former world top-ten players and major champions who saw their careers fall apart after too much tinkering.

If it's so hard for the world's number-one golfer to learn how to hit a draw, imagine the difficulty for an average guy like me to add a new shot and hit it consistently. Kaymer's experience shows just how fragile some hard-earned skills can be, and how even the best athletes in the world can struggle to learn a new skill. Anything that an athlete and his support team can do to speed the process—to optimize learning and retention—can have a huge effect on competitive outcomes. But amazingly, even the best performers and coaches often miss the chance to leverage the science of learning.

Practicing for Performance

John Kessel was a midthirties American playing professional volleyball for the Alessandria Volleyball Club in Italy in 1983. His team was on the cusp of a huge victory over their archrival, Moncalieri, which Alessandria hadn't beaten in seventeen years. "It was an away match," says Kessel. "There were thousands of fans there." Alessandria was leading two games to none and 13-5 in the third game, and serving. First one Alessandria player served the ball into the net, then another. They seemed to be falling apart as Kessel stepped up to serve.

Kessel's coach shouted instructions to him: *"Non batte la palla nella rete!"* Don't serve the ball into the net!

"Va bene," replied Kessel.

"Then I turned around and served the ball backwards into the crowd as high as I could," says Kessel, laughing at the memory. "We went on to win the match; they overturned our bus and we had to sit in the locker room for, like, two hours while the other team's fans rioted because they'd lost to us. And my coach says, *'Che cazzo fai?'* which translates as 'What the fuck did you just do?'

"I said, 'I think you told me not to serve into the net, and I didn't.'"

Kessel then asked his teammates what they thought about when the coach told them not to serve into the net, and they said all they could think about was, of course, serving into the net. It's like a golfer who stands on the tee looking at a huge fairway with water down one side and thinks, "Don't hit it into the water." It's a recipe for doing just the thing you're trying to avoid.

It's a good story, one that Kessel uses to illustrate the importance of how the best coaches use positive feedback for their players rather than negative. But it's also a good illustration of the mind-set at USA Volleyball, where Kessel is director of sport development. It's a mind-set that challenges received wisdom, pushes athletes and coaches to improve constantly, and has delivered surprising success.

I love indoor volleyball. While beach volleyball has become a telegenic event featuring tall athletes in skimpy clothes at the past few Olympics, the indoor game is a hell of a sport. It features speed, coordination, power, tactics, skill—it doesn't get as much credit as it should in the United States, especially for male players.

Volleyball is actually a hugely popular sport for women in the United States. According to the National Federation of State High School Associations participation survey, there are about 420,000 girls playing high school volleyball in the U.S., making it the third-biggest sport in terms of female participation, trailing only track and field and basketball (and still well ahead of soccer). When it comes time to compete at the next level, there are more than twelve thousand college scholarships available to female volleyball players.

Compare that with the men's game. There are about 160 total scholarships available at the college level for male volleyball players, as there are just twenty-three programs at the Division I level for men, compared with 329 for women. That's not much of a chance for a male volleyball player to compete at a level beyond a recreational one in the U.S.

But then you look at the results from the Olympic Games. Since the sport entered the Olympic program in 1964, the American women have never won a gold medal—they've won three silver and one bronze. The U.S. men, on the other hand—even with a much, much smaller population of players to choose from—have won three gold medals and one bronze. What gives? How have the U.S. men topped the women when there are far fewer players? You'd assume that more players creates a better chance of finding great players and working to improve them.

Kessel points to one key factor that has allowed the U.S. men to excel. It's all about how they teach and coach the players.

Traditionally, volleyball has been coached in a very structured way. The model was established early in the game's Olympic history. The first women's gold medal was won by Japan in 1964, and the team was a marvel of technical ability, with great ball control and skill. The Japanese players would spend long chunks of practice first working on one skill, like passing, and then another on serving, another on setting. As athletes would master one part of the skill, they would add more parts, in a steady progression.

The U.S. Men's National Volleyball Team followed that regimented Asian style of practice until a coach named Carl McGown took over the program in 1973. The U.S. men were nowhere in terms of competitiveness at the international level—the team had finished a dismal eighteenth in the 1970 World Championships and didn't even qualify for the 1972 Olympics.

McGown had been an all-American volleyball player during college, and then went on to earn a PhD at the University of Oregon. But the crucial thing was what his doctorate was in: motor skill learning. Motor learning is a branch of science that deals with how we best learn new physical (or motor) skills, and in his studies, McGown learned that, well, almost everything that volleyball coaches were doing was wrong.

The structured practice that Japan used in volleyball has a couple of key characteristics. In the language of motor learning, it's both "blocked" practice and "part" training. The players would work on just one small facet of a skill (that's the "part" part), and they'd do nothing but work on that until they moved to the next task (that's the "blocked"). McGown knew that both of those types of learning had been called into question in the motor learning literature, in favor of "random" practice and "whole" training.

John Kessel has a story he likes to use to demonstrate the gulf between how we learn when we're left on our own and how coaches usually try to teach athletes. "Did your parents hire a bike riding coach to teach you?" he asks. "Did they send you to a bike riding summer camp or put you through bike riding drills? Did they ask you to do progressions of various skills, like twenty pedal strokes with your right foot and then twenty with your left?" When we learn how to ride a bike, most of us get a couple of instructions (like, "Keep pedaling!") and then are left to figure it out on our own as we do it.

This is random practice—learning to cope with what comes along as we go, organically, rather than in a set, drilled environment. And it's whole training—learning the entire skill of riding the bike rather than its components.

It's not surprising that coaches are drawn to blocked practice. After all, it's easier to set up and run. But it's simply not as good for the final outcome, which was demonstrated by a clever experiment published in 1979 by John Shea and Robyn Morgan at the University of Colorado. They set out to show the differences between blocked and random practice in the acquisition of a new skill. They took two groups of students and set out to teach them three different tasks (in this case, moving their hands and arms in a certain way when they saw a light signal). One group did all of their learning and practice in a blocked way—they first learned movement A, then movement B, and then movement C. The other group did random practice, in which the patterns never repeated more than once.

When it came to learning and practicing the skill, the group that did the blocked practice performed better than the group that did random practice. They knew what was coming next after a certain period of time, and could complete the actions more quickly. The random group struggled comparatively, taking longer to complete each action.

But here's the rub: When the students were then tested on their *retention* of what they had learned, the random group was much, much better than the blocked group. Even after just a ten-minute break, the group that had learned and practiced in a random manner performed better than the blocked group, even if the retention trial was done in a blocked manner.

Blocked practice might make it easier to learn, but you actually develop a skill over the long term more effectively in random practice.

Richard Schmidt, one of the leading researchers in motor learning, put it this way when he was talking to a group of volleyball coaches: "Are you practicing for practice, or are you practicing for performance?" Of course, the answer is that you should be practicing for performance—for the competitive situation—but often coaches and athletes overlook that fact, because blocked practice feels more polished. "Practice shouldn't look good," says Kessel. It's a time to learn, not to show off.

That's a lesson that Peter Vint, a sports scientist at the U.S. Olympic Committee, is trying to teach to other sports. To start to incorporate these ideas, says Vint, "you have to take a leap of faith. If you adopt an approach that introduces more variation and stress in practice, performance in practice will suffer." And that's scary to coaches and owners.

But some teams have taken that leap. Vint recently worked with one of the top teams in the NBA to help them implement changes to their routines. For instance, the team had been practicing free throw shooting the way nearly every basketball team in the history of the game has—by taking a specific time to shoot free throw after free throw. But of course, in a game, you never shoot more than two (or on rare occasions, three) free throws in a row, and you don't shoot them in a rested state—you're always having to ramp down from game action. So instead of the standard blocked free throw practice, now the players never take more than two free throws in a row, and they always make sure the free throws are mixed in with other high-intensity activities, so their heart rate isn't less than 85 percent of their max. By making the practice random and realistic, the team hopes to ingrain the habits that will help when it comes to in-game performance. "You're mimicking the realities of the competitive situation," says Vint.

The team has also started something called Bad Pass Friday. Usually when players practice shooting, coaches throw them nice passes right where they like to catch the ball. But in a game, when there's a defender trying to disrupt the play at every opportunity, passes rarely arrive perfectly before you shoot. So on Bad Pass Fridays, coaches throw passes that are still catchable, but just not *easy* to catch.

After one of these drills, a top shooter in the league told one of the coaches, "Coach, that was a lot harder." As Vint says in describing the drill, when you have players at the NBA level being really challenged by a drill in practice—when you have them actually improving through practice—that's a pretty special thing.

Bad Pass Friday hits on another keystone of motor learning theory, which is the importance of specificity. The best way to learn for performance is to do something that's as similar to the actual performance environment as you can. The way Kessel puts it is, "The game teaches the game." Rather than having players run through traditional drills, the goal is to create situations that are as close to the actual game as possible, because that's the best way to learn about the actual situations you'll face when playing for keeps.

"I read about NFL quarterbacks at practice, talking about making fifty of the same throw on the same pass route, with the same pace on the ball and the same conditions, so they can 'get in the groove,'" says Vint. "But you'll never be in the groove under real-world pressure."

Moving Pictures

Every time Olympic gold medal skier Ted Ligety steps into his bindings for a training run, there's more than a stopwatch capturing his performance. Each time the gymnasts on the U.S. Olympic team train on the pommel horse or the still rings, there's not just a coach watching. Whenever the ball is snapped by an NFL team—any team, in practice or a game—there's a silent observer that grabs the play for posterity. In today's world of sports, there is almost always a video camera. And for good reason.

In team sports, there are the obvious tactical reasons to tape games. Coaches have been breaking down "film" for decades, looking for the plays and patterns that their opponents fall into, trying to find a tactical advantage. Individual players study the tendencies of their opponents—does a guy like to drive right or left to the basket? How does he defend a pick-and-roll? New recording and organization systems have made this work much easier in recent years: If I'm a baseball player and I want to see all the at-bats I've had against a certain pitcher in my career, I can get the entire collection on my iPad in minutes.

But beyond that sort of tactical analysis, video has become an essential tool in learning and perfecting skills. "I'm a visual learner," says Ligety. "I'm able to take video and pick out the fundamental of the movement that I'm trying to work on. It's easier for me than being told what to do."

It's not an exaggeration to say that every athlete and coach I have spoken to in my research pointed out the importance of video for feedback in skill acquisition. Scientists have started to explore the seeming affinity between athletes and visual presentation of information, and they're finding that compared to the rest of us, athletes are amazing visual processors.

Jocelyn Faubert, at the University of Montreal, set out to test the ability of professional athletes to process a complicated visual scene, as compared to elite amateur athletes and nonathletes. There have been plenty of studies that show athletes are very good at visually processing scenes related to their sport, but the test Faubert used wasn't sports specific, but rather a pattern of circles that the subjects had to track over time.

Perhaps unsurprisingly, the professional athletes from the National Hockey League and soccer's English Premier League outperformed NCAA-level athletes and nonathletes from the very first trial. After all, the pros rely on visual pattern recognition every day. But more surprisingly, they also improved much more rapidly than the other groups as they practiced the task. As Dr. Faubert said in a release, "We were not surprised to see athletes' initial scores correlate with levels of sporting ability, but we were amazed to discover learning capacity to be such a pivotal marker. It's fascinating data." The pros didn't just start out better; they also got even better faster. Faubert's results suggest that it's not just the sport-specific visual abilities that help those athletes but also an inherent ability to improve more quickly.

That visual fluency might explain why video is an essential tool to help athletes bridge the gap between their mental image of a performance and the reality of it. You and I can tape our golf swing or running form—and in fact, even nonelites can get great feedback from looking at those tapes. But the best athletes can extract more information from that video, and are more adept at translating what they see into what they feel during an event.

"One of the most frustrating things as an athlete is doing something and thinking that you know what you look like," says decathlon world record holder Ashton Eaton. "But then someone says, 'That's not what you look like at all.' Video means I can see the things I'm doing wrong even though I feel like I'm doing them right."

That's why every single training run that Ligety does is recorded. They use the video not only to help offer feedback on what he's doing but also to isolate the performance of his skis. Head, Ligety's ski manufacturer, takes high-speed video focused on the ski rather than Ligety. They can then analyze how the ski is running through the snow.

Ligety also uses the GoPro video camera, which was invented by Nick Woodman, who was trying to find a way to take good pictures of himself when he was surfing. The company's rugged, waterproof cameras have become a constant companion to many competitors, especially in action and extreme sports. The small unit, a rectangle that's only a couple of inches wide, can be mounted to a helmet or a bike—or nearly anything you can think of—and then capture high-def video with the press of a button.

Ligety has spent some time building himself a special rig for his Go-Pro, which extends two feet behind his head while he's skiing, focusing the camera on his body and skis. "You get a really cool view of the entire run the whole way down," says Ligety. "You get a good sense of how your body is lining up at the top of the turn, and where the pressure is being applied to the ski in the turn."

In the videos, Ligety's body stays centered under the camera, while his skis turn left and right across it. The effect is hypnotic, but for Ligety, it's an important tool to try and find a winning edge.

Hammer Time

The seemingly simple insight that the best way to get better at a particular sport is to practice it was the competitive advantage for one of the greatest coaches in any sport's history: Dr. Anatoly Bondarchuk. Bondarchuk became the 1972 Olympic gold medalist in the hammer throw for the Soviet Union, while also completing his doctorate in pedagogical science at Kyiv University. In 1976, he won the bronze medal in the hammer throw—and was the coach of the gold and silver medalists. The top three hammer throwers in the 1980 Olympic Games? The 1988 games? The 1992 games? All Bondarchuk's athletes. Only the Soviet bloc boycott of the 1984 Los Angeles Olympics could stop Bondarchuk's juggernaut.

Bondarchuk had long been a shadowy figure to athletes in the West, the mysterious coach who churned out champion throwers like clockwork. But in 2004, his daughter was set to move to Canada, so Bondarchuk e-mailed the small Kamloops Track and Field Club in British Columbia, Canada. Would they be interested in hiring him as a coach? One member of Kamloops was Dylan Armstrong, a top Canadian thrower, and they happily accepted. It was like Bill Belichick calling an amateur football team in Russia and asking if they'd like him to give up his job with the New England Patriots to take over their team.

After his arrival in Canada, North American throwers flocked to the club in the town of eighty-five thousand people. There, they discovered that there was no crazy Soviet secret that had allowed Bondarchuk to deliver all those champions. Instead, there were a couple of key principles that guided the construction of the athletes' training programs. The foremost among these was Bondarchuk's concept of "special strength."

Martin Bingisser, a Swiss national champion in the hammer throw who trains with Bondarchuk, explained the concept in an article for the journal *Modern Athlete & Coach*:

> Strength training is a great short-term investment. If a young athlete adds strength, their results will quickly improve. How-

ever, over the long term, the results plateau. Bondarchuk's research has compared two groups of athletes: one group focused on lifting heavy and the other focused more on throwing-specific exercises. While the first group made faster progress in the beginning, the second group inevitably surpassed the first group after about four years. In other words, lifting will only take you so far.

Over the course of his coaching career, Bondarchuk has tracked the correlation between an athlete's performance in various strength exercises and his performance in the competitive event. As Bingisser points out, simply being strong is an advantage at the lower levels of competition. Even at a very good level—say, a sixty- to sixty-five-meter hammer throw—there's some correlation between an athlete's performance at a weight room exercise like the squat and his hammer throw results. But once you get to where you need to be to win on the world level (seventy-five to eighty meters), those correlations essentially disappear. If you tell me how much weight an athlete can lift at the high school or college level, I'll have a good guess at how he'll finish in the hammer throw. But if you give me those stats for someone at the world championship level, I'll have no idea. Once you get to a certain threshold, differences in what you can lift in the weight room don't mean squat.

At that superelite level, the only training exercises that retain significant correlation with competitive results are . . . throwing the hammer. Like Kessel says, "The game teaches the game." You can see the connections between Bondarchuk's theories and motor learning theory. It seems obvious when you state it, but all around the world, thousands of coaches and athletes miss the correlation, focusing instead on hours in the weight room rather than spending time on the sport itself. The game teaches the game—not just tactically, like in team sports, but physically as well. Work in the gym is important, but as we've seen from coaches like Bondarchuk (as well as the work that people like Phil Wagner at Sparta Performance Science have done), it's only one tool in support of the athlete's performance in a real setting, and not an end in and of itself.

Small Games, Big Impact

For team sports, there's a step beyond using the game itself to teach the game, and that's using what are sometimes called "small-sided" games. The idea is to replicate the sport using fewer players than the standard game, so that each player has to make more tactical decisions and touch the ball more often. It's fairly common to see three-on-three basketball tournaments, which give players an opportunity for this type of experience. Going even further, basketball has what may be the ultimate small-sided game, in one-on-one. (Volleyball coach Kessel says he's jealous that one-on-one in basketball is useful, since you really can't play volleyball one-on-one. You *can* play two-on-two—that's part of why beach volleyball has become so popular.)

Daniel Coyle, in his book *The Talent Code*, explores the power of an indoor version of soccer called futsal. Futsal is a five-on-five game played on a small court with a heavier ball that requires better control than a regular soccer ball. "Futsal compresses soccer's essential skills into a small box; it places players inside the deep practice zone, making and correcting errors, constantly generating solutions to vivid problems," Coyle writes. "Players touching the ball 600 percent more often learn far faster, without realizing it, than they would in the vast, bouncy expanse of the outdoor game." Coyle meaningfully points out that futsal is a superpopular sport in Brazil, a country known for its soccer players' brilliant ball skills. The list of soccer superstars who played futsal as kids in Brazil is overwhelming, including Ronaldo, Ronaldinho, Neymar, and Pele. Outside of Brazil, three of the best players in the world today—Argentina's Lionel Messi, Portugal's Cristiano Ronaldo, and Spain's Xavi—all played the small-sided version of soccer. "During my childhood in Portugal, all we played was futsal," Cristiano Ronaldo told FIFA.com. "The small playing area helped me improve my close control, and whenever I played futsal I felt free. If it wasn't for futsal, I wouldn't be the player I am today."

The advantages of small-sided games have now come to American football, which has always had a difficult problem to overcome: The physical demands of the game and the short season make it hard for players to get as much exposure to competitive situations as you would like them to have. The solution has turned out to be seven-on-seven football, which has taken the country by storm in the past five years at the high school level—nowhere more than Texas.

It's basically touch football. There are two twenty-minute halves, and the game's played on a forty-five-yard-long field. There's no blocking, no pass rush, and no running the ball. Instead, it's a quarterback, a center, and five receivers on each play, trying to move the ball against seven defenders. After the ball is snapped, the QB has four seconds to throw before the play is whistled dead, and receivers are down by touch after they make the catch; you can get a first down by advancing the ball fifteen yards.

The fast-paced nature of the game means that a quarterback might throw seventy or more passes in a contest, while receivers might find themselves targeted more than ten times apiece. And then when you realize that teams play four or five games a day during a big tournament, that's an amazing amount of football to play, compressing those skills (as Coyle says about futsal) into a small box.

Since the first Texas state championship was organized in 1998, the seven-on-seven game has boomed. The first tournament had only thirty-two high school teams; the 2013 version featured 128 teams in two divisions (roughly 2,000 players). To even get to that tournament, teams had to make it through one of thirty-two qualifying tournaments around Texas.

There's no denying the effects for players, especially quarterbacks. In the 2012 NFL draft, three of the first eight players chosen were QBs: first overall pick Andrew Luck for the Indianapolis Colts, second pick Robert Griffin III for the Washington Redskins, and eighth pick Ryan Tannehill for the Miami Dolphins. All three grew up in Texas, and all three grew up playing seven-on-seven. Before the draft, Tannehill told *USA Today* how important the small-sided version of football was to him. "It was huge in developing my passing touch and accuracy," he said. "It's just the repetitions of throwing the ball every day during the summer. It gets you throwing every route that's out there—the out routes, the in routes, the deep routes, the short routes. You get to see all types of coverage—man coverage, zone coverage, combo coverage. You see everything in seven-on-seven. It's a great tool in developing quarterbacks."

Madden Men

When you want to teach someone a complex, dangerous skill, computer simulation is a powerful tool. Airline pilots spend hours in simulators that re-create the cockpit of a plane down to the individual knobs and switches, practicing not just routine tasks but also emergency maneuvers and procedures that they (hopefully) will never face in the air. The ability to do these recoveries from simulated near-disaster offers them a wealth of experience to draw upon without the cost and danger of trying them in an actual airplane.

Meanwhile, simulation has become such a large part of medical education that there are now entire textbooks dedicated to the field, from physical simulators that teach surgeons the skills required to perform a particular operation to virtual reality setups that help guide students through human anatomy. Some researchers have even built simulations in the virtual world Second Life to assess students' diagnostic and communication skills, requiring the residents to assess virtual patients suffering from various maladies like gastrointestinal bleeding and bowel obstruction. "The way we learn in residency currently has been called 'training by chance,' because you don't know what is coming through the door next," said Dr. Rajesh Aggarwal, who helped build the system, in an American College of Surgeons press release. "What we are doing is taking the chance encounters out of the way residents learn and forming a structured approach to training."

Simulation allows you to do a couple of things. First, it gives you the chance for many more repetitions of a skill than you might be able to get without them. Thankfully, there aren't enough appendectomies or emergency landings required in the real world to give people as much practice as they can get with a simulator. And the simulation, importantly, allows you to avoid the consequences of mistakes from a novice surgeon or pilot. A failure in a simulator might lead to some disappointment and valuable feedback; failure in the real world can lead to much more dire outcomes.

Football is struggling with its own dire outcomes. The sport is under fire at every level, from Pop Warner to the NFL, due to our increased awareness and understanding of the dangers of concussions and other head injuries. The NFL is still, by far, the most popular sports league in America, but the game is shrinking at the youth level—as of this writing, the number of participants in tackle football has decreased each of the past five years, according to the Sports and Fitness Industry Association, dropping from 7.9 million players in 2007 to 6.2 million in 2012.

But just like for pilots and doctors, there's a simulator available for football players that allows them to play the game without encountering the physical dangers and consequences of the actual sport. It's an incredibly granular re-creation of the game, with a level of realism that can feel almost indistinguishable from football on TV. For a product of such sophistication, it's quite affordable: You can grab it at your local mall for about fifty bucks.

I'm, of course, talking about *Madden NFL* football, the video game series published by Electronic Arts. Since the company released its first version of the game twenty-five years ago for the Apple II computer, it's become one of the best-selling video game franchises of all time, selling in excess of 100 million copies and raking in billions for EA. But I'd argue that Madden has had a profound effect on how the game is played in the real world, by tapping into the power of simulation to teach essential skills.

No pro football player in the game today has lived in a world without *Madden NFL*, and any player is exceptionally likely to have spent a vast part of his youth—hundreds, if not thousands, of hours—playing it. That's a massive amount of practice, and there's a very real possibility that all those hours directly translate to the gridiron. Football isn't just a physical game; it's a game of visual pattern recognition and memorization as well. Of course, you need to be strong and fit and fast to play professional football, but you also need to have a very specific kind of knowledge and intelligence. You need to be able, if you're an offensive player, to look at a defensive alignment and pick out the key opponents whose actions will expose the defense's intentions, whether it's a safety creeping up to the line of scrimmage to stop the run or a linebacker dropping back to defend the pass. If you're a defender, you need to be able to view an offensive formation and adjust your coverage to match what you see.

That visual training is what's happening, even subconsciously, every time a kid plays *Madden*. Your average thirteen-year-old fanatic has likely seen more football plays develop visually than a Hall of Famer could over the course of a pre–video game career. Gamers are gaining cognitive, if not physical, experience every time they turn on their Xbox or PlayStation. They might not all have (or grow into) the physical size or skills to play pro football, but they're developing a sophisticated visual understanding of the sport. And for players who do have the ability to play in the NFL, that understanding can be a huge boost.

In the 1990s, Tim Grunhard was an All-Pro playing center for the Kansas City Chiefs. He was the offensive line coach at the University of Kansas, but before that, he was a high school coach who used *Madden* to help his players learn the game. "These games nowadays are just so technically sound that they're a learning tool," Grunhard told *WIRED*. "Back when I was playing football, we didn't realize what a near or a far formation was, we didn't really understand what trips meant, we didn't understand what cover 2, cover 3, and cover zero meant." These are all types of formations and defensive coverage that his players were learning through the video game. Plus, they were acquiring the skills to see how plays develop. "It just seemed to help out," Grunhard said. "The kids understood where the counter play or power play was going to open up. Or the middle linebacker lining up for a blitz—where the gaps were going to open up."

That understanding has allowed the game to become increasingly fast and complicated. The latest craze that's swept the game, from the professional ranks down to high schools, is an up-tempo offensive style called the read option. It's an offense that requires the quarterback to "read" various defenders and make instant decisions about what action to take with the ball—hand it off to a running back, keep it and run himself, or throw a play action pass. When executed correctly, it's devastatingly effective, but it requires not only an extremely athletic quarterback, but one who can make those decisions correctly and quickly. The pace is fast, the action frenetic—watching a team like the University of Oregon execute it feels like, well, watching a video game.

Madden has become an interesting technological tool for teaching football, but the importance of an athlete's tools doesn't end with the virtual. Some of the most important are the things that they take onto the field of play when the competition starts.

7

TOOLS OF THE TRADE

From Skis to Golf Balls, the Hidden Consequences of Sports Gear

Imagine that you are a member of the U.S. Women's National Soccer Team. You're part of one of the dominant programs in international sports, a team that's won three consecutive Olympic golds and two World Cup titles. But say one day you wake up to an unexpected announcement: Soccer officials have decided to make the goal bigger. Instead of 24 feet by 8 feet, it's now going to be 26 feet by 9 feet. If you were a striker, you'd be thrilled, of course. You've suddenly got a target that's grown from 192 square feet to 234 square feet. If you were a goalie or defender, however, you'd have a very different reaction—with a 22 percent bigger goal, your job just got much harder. But one thing would be clear: The game simply wouldn't be the same.

What if Major League Baseball decided that the baseball would be an inch larger in diameter? What would the impact on the game be? Would offense go up, since the larger ball would be easier to make contact with? Or would it go down, because a bigger projectile wouldn't fly as far once it was hit? Would pitchers with big hands have an advantage because they could still make it curve? A seemingly small change to the equipment would have a profound effect on the sport.

That's basically what happened in 2012, when the Fédération Internationale de Ski decided to change the sport of alpine skiing, and most radically, the giant slalom event. Since the 1990s, ski racers had used what were at first called shaped skis, that is, skis that were shorter and had more sidecut (the arc on the edges of each ski) than traditional straight skis. Shaped skis are much, much easier to turn, because the increased sidecut and shorter length mean that when a skier rolls his leg to put the ski on edge, it naturally tries to carve a turn.

The technology revolutionized the sport, both at the World Cup level and at the recreational level. Millions of skiers experienced their first-ever carved turns, rather than sliding through them. And racers were able to ski in a very different, more aggressive style. They could leave their turns until later on the course because of the skis' ability to carve so much more sharply. And since the racers were strong enough to flex the skis themselves, they got a level of rebound out of them at the end of each turn, helping to almost rocket them into the next gate.

The technical limits on the shape and design of skis in racing are set by the FIS. The federation was concerned by what it saw as an epidemic of injuries in the sport, and decided to look at the skis themselves to see if they could be modified to make the sport safer. The federation ran tests with scientists from the University of Salzburg, in Austria, putting recently retired World Cup skiers on various prototype skis and measuring the forces generated as they ran through the gates on a racecourse. The idea was that by reducing the forces generated in a skier's body, you would reduce the risk of injury, especially to the knees, which deal with tremendous stress during a competitive ski run. Through this testing, the FIS arrived at what it viewed as a solution—make the ski less aggressive in terms of length and shape.

Prior to the new rules, giant slalom skis were required to be at least 185 cm long, and the radius of the sidecut was required to be twenty-seven meters. The new rules required that the skis now be at least 195 cm long, and that the radius, which measures the arc the ski can carve, be increased to thirty-five meters. After two decades of shorter skis with more sidecut, racers would have to ski on longer, straighter boards, a type of ski that most of them had never even tried.

Immediately, there were howls of protest from manufacturers and the skiers on the World Cup tour. One of the most outspoken opponents was American skier Ted Ligety. Ligety, whom we met in the previous chapter, is one of the best skiers the U.S. has ever produced—a surprise gold medalist at the 2006 Winter Olympics in Turin, he continued on to an excellent career, mostly in the giant slalom. He's won twenty-two World Cup races, two World Championship golds and one bronze medal, and four World Cup giant slalom titles.

To say that Ligety wasn't happy with the new ski regulations would be a gross understatement. After the rule change was announced, he took to his website, where he posted several pieces outlining his objections to the new rules. "Today I finally had the chance to try a prototype of the . . . skis and quite frankly they suck," Ligety wrote. Rather than the carved turns he was used to, the new skis seemed to skid around corners.

Ligety was by no means alone. Eventually, fifty-six of the top sixty skiers in the world signed a petition stating their opposition to the new rules, but the FIS was unmoved. The rules took effect for the first race of the 2012–13 season, a giant slalom in late October on the Rettenbach glacier, in Sölden, Austria. Ligety won by 2.75 seconds, the largest margin of victory in a World Cup GS in thirty-four years. He won the next World Cup GS race, at Beaver Creek in Colorado. After a third place at Val d'Isère, in France, Ligety won again and again. For the season, he won an amazing six of the eight giant slalom races held, capturing the World Cup title for a fifth time. He added three gold medals in the 2013 World Championships, capping by far his most successful season ever.

"My technique really matches up with the new skis," Ligety tells me. "I start my turns earlier and finish them later than a lot of other guys, and I'm really clean as I go from edge to edge. I don't generate as much rebound out of the ski as some guys." The man who hated the new skis so much was nearly impossible to beat when he raced on them.

The changes in the results were striking. Some skiers, like Ligety, thrived with the technique the new skis required. But other competitors found their results in a free fall. Take the case of Italian skier Massimiliano Blardone. In 2012, Blardone was third in the season-long World Cup for giant slalom. In 2013, on the new skis, he was thirteenth, crashing in three of the eight races, and earning half as many points in each race he entered. "He used to take a straight line and chop off the top of his turn," says Ligety when asked to diagnose Blardone's struggles. "He can't do that anymore on the new skis."

Ligety says that he doesn't like that the FIS can make equipment changes that favor one skier and hurt another—even though the changes have favored him. But what's happened in skiing illustrates something that we often forget. The tools we use to play our sports—the equipment and venues—can have a profound impact on the sport itself. They might be intentional or unforeseen, but every decision about the technology and equipment in a sport has its consequences.

From Formula 1 to Foam Bricks

There's not a sport on earth that doesn't have some sort of technological component—think of all sports as points on a continuum that ranges from events whose outcomes are determined primarily by technology and equipment to those that are determined primarily by human performance.

At one end, you'd find Formula 1 auto racing. This isn't to say that Formula 1 drivers aren't great athletes, because they are. Consider, if you will, the stress placed on the neck of an F1 driver over the course of a two-hour race. In each turn, the driver has to deal with forces on the order of 5 g's—five times the normal force of gravity. Even keeping your head still, which you'd like to do so you can, you know, drive the car, is a battle. But he's also wearing a helmet and a neck restraint that adds about four pounds to his head. Multiply that by the g-forces and suddenly F1 drivers need the neck strength to keep a seventy-pound weight stable so they can see where they're trying to go. And they need to do that through the hundreds of corners that they'll encounter in a typical Grand Prix.

To build that strength, drivers use a special rig that's basically a helmet with pulleys and weights attached. They have to resist the force of the weights, developing their neck strength. How strong do their necks get? F1 World Champion driver Fernando Alonso likes to crack walnuts using his neck muscles. (There's an awesome video clip on YouTube showing Alonso doing this party trick. It's worth watching.)

The neck strength isn't all. The brake pedal on an F1 car requires a huge amount of force to activate, upwards of one hundred pounds of pressure. So, again, over the course of a Grand Prix, drivers are doing hundreds of virtual leg press repetitions (with one leg) at more than one hundred pounds. Not easy. They need the cardiovascular fitness to keep their muscles functioning at high g-forces, often in extremely hot conditions, as they're crammed into tiny cockpits near all the mechanics of the engine and exhaust systems. Oh, and no power steering, so they need very strong arms to not only keep them extended forward for two hours (plus fighting g-forces) but also wrench their cars around corners.

All this is to say that F1 drivers are really fit—many of them do cycling races or triathlons for fun as well as training. But the competitive result of an F1 race is most influenced by the car itself. While a great driver might be able to maneuver his car from the middle of the pack closer to the front, the best driver in the world can't win Grand Prix in a substandard car. There have been years when certain cars have been so much better than the competitors' that it seemed to make the matter of who drove irrelevant. Formula 1 is, at its heart, a competition between engineers.

(Interestingly, Formula 1 cars aren't as tricked out as they could be, or as they have been in the past. Over the past decade, Formula 1 has banned both traction-control technologies and antilock brakes in an effort to make the cars' performance more reliant on the drivers' skill than the engineers'. This leads to the very strange situation in which my Volkswagen Jetta is, in some ways, more technologically advanced than a race car worth more than $2.5 million.)

Let's contrast that with one of the least technologically influenced sports, Olympic wrestling. There's not much in the way of equipment: a mat, a singlet, maybe headgear to protect the wrestler's ears, a pair of shoes.

Ah, the shoes. Even in such a simple sport, a lot of engineering gets packed into a product that's less than eight ounces. For instance, there's the Asics Omniflex-Pursuit wrestling shoe, which does away with much of the traditional upper, paring away material in favor of a series of straps over a compression sock liner. The idea is to give the wrestler a feel that's as close as possible to barefoot, but with as much traction as possible on the mat. One pair will set you back a cool $160.

But technology can have an effect on a sport outside of what it does for equipment. Wrestling allows for video replay on scoring decisions made by officials. The system works a little bit like the NFL's: If a wrestler and his coaches want to review the call made by the referee during a match, they ask for a challenge. Unlike in the NFL, they don't throw a flag; they throw a foam brick (not sure why, but the phrase "challenge brick" is pretty great).

The impetus for the current system came after the 2008 Olympics, when Swedish wrestler Ara Abrahamian lost in the semifinals of the 84 kg Greco-Roman competition. Abrahamian and his coaches disagreed with the ruling of the officials on a crucial point in the match, and indeed, the video replay seemed to show that Abrahamian was correct. But there was no mechanism in the rules to view the replay or to change the decision. Abrahamian would later take his bronze medal and leave it in the middle of the mat after the medal ceremony, an action that caused the International Olympic Committee to suspend him, decrying his lack of respect for "the spirit of fair play." He was stripped of his medal and kicked out of the games. But later, in an appeal to the Court of Arbitration for Sport—essentially the Supreme Court of the international sports world—Abrahamian won a judgment stipulating that wrestling officials must have some sort of review and challenge system in place. Hence, the challenge brick.

The system's not without its problems, including the possibilities of an athlete who lodges a challenge having points taken away through the review process. But it has had a large effect on the sport, and taken away a little of the ambiguity that has always existed in the judging of the fast-moving action of a wrestling match.

Improving by Leaps and Bounds

So every sport has its technological component. Some sports have been completely changed by a modification in technology or equipment. For example, let's look at the world record for the men's pole vault over the history of the event, plotting the height of each jump in meters.

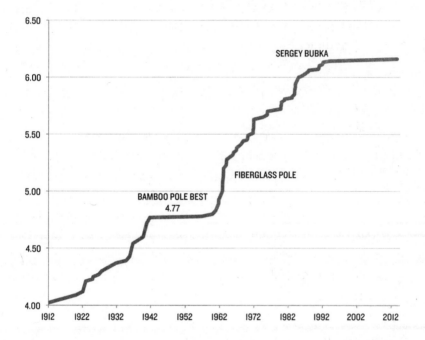

You can see a relatively steady progression in the world record from the first IAAF-ratified record of 4.02 meters in June 1912. That progression continues until the amazingly named Cornelius Warmerdam, of the United States, set the mark at 4.77 meters (that's 15 feet, 7¾ inches) at a meet in Modesto, California, in 1942. What all those records had in common was that they were set using a bamboo pole, which had minimal flexibility. Essentially, the vaulter used the pole more like a lever to launch off the ground. Warmerdam's record lasted for fifteen years; it appeared he had effectively maxed out the possibilities of the bamboo pole.

The next two world records were set in 1957 and 1960, by jumpers who used an aluminum pole and a steel pole, respectively. Those poles still weren't very flexible, but their construction was a little lighter than bamboo, allowing the athlete to get a touch more speed down the runway.

But then we see that nearly straight vertical line on the record progression, starting on May 20, 1961, when American George Davies became the first vaulter to set a world record using a fiberglass pole. Within two years, the record went from Davies's 4.83 meter vault to John Pennel's (also American) 5.20 meters. In two years, the record spiked 37 centimeters (more than a foot). The previous 37 centimeters of improvement in the pole vault world record had taken twenty-four years to accomplish.

The difference, of course, was the way the fiberglass (and later, carbon fiber and other composite material) poles worked. As the pole was planted, it flexed, absorbing some of the energy accumulated by the vaulter during his run-up. As the vaulter began to fly through the air, the pole uncoiled like a spring, returning the energy to the vaulter, allowing him to go much, much higher than he would have with earlier poles. Over the course of his amazing career, Ukrainian vaulter Sergey Bubka set the indoor and outdoor world record thirty-five times, boosting it from 5.81 meters to 6.15 meters. That final record, which he set in 1993, was finally broken by French vaulter Renaud Lavillenie in early 2014, after standing for twenty-one years. You might say that Bubka is the Cornelius Warmerdam of the pole vault's composite age.

Golf has dealt with similar technological changes. Back in 1980, Dan Pohl was the king of the long hitters in golf. He wasn't a big guy, standing at 5 feet 11 inches and weighing 175 pounds. But he could really move a golf ball off the tee. By using a long, wide backswing with a late loading of the club, he generated tremendous clubhead speed. That allowed him to lead the PGA Tour in driving distance that year, outhitting noted power players like Jack Nicklaus. Pohl averaged 274.1 yards per drive, while the tour average for the season was 256.9 yards.

Fast-forward thirty-three years. If Pohl had posted those same numbers—remember, the best driving numbers on tour in 1980—in 2013, he would have ranked 175 in driving distance on the PGA Tour.

Only *six* players would have finished behind him. The tour leader in 2013 was Luke List, who averaged 306.3 yards a drive, a full 32.2 yards longer than Pohl in 1980. The Tour average was 287.2 yards, an increase of 30.3 yards.

Like the pole vault, once you start to plot out those increases over time, it's very easy to see the inflection points where driving distance starts to increase radically. Here's the average driving distance in yards on tour since 1980:

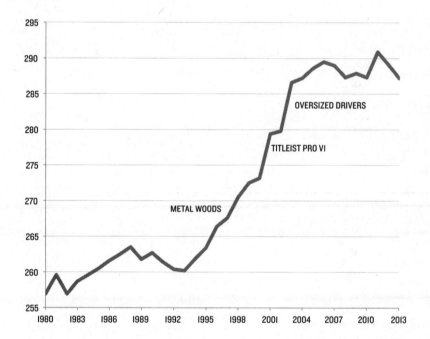

You can see that there are three points where things really change. From 1980 to 1993, there's a gradual improvement in driving distance, with some fluctuation, to 260 yards or so. But in 1993, a steady increase begins to build, with a thirteen-yard gain in average distance from 1993 to 2000. In 1993, Bernhard Langer won the Masters, one of golf's four major tournaments, using a driver made of persimmon, rather than one of the new generation of metal "woods" that had been slowly infiltrating the game. He would be the last player to win a major with an actual wooden wood—by 1997, Davis Love III retired his persimmon driver and old-school woods left the tour for good. Meanwhile, metal drivers were becoming better and better, leading to that steady improvement.

And then things got a little nuts. The next year, 2001, average driving distance leapt six yards in a single season. There was a very clear reason for that huge jump—the introduction of what might be the single most influential product in the history of any sport: the Titleist Pro V1 golf ball.

For decades, top golfers had all played with balls constructed in the same way: A liquid-filled rubber core was wound with thin rubber thread, building the ball up to the correct diameter as if it was a ball of yarn. This was covered with balata, a type of rubber harvested from a tropical tree called the bully tree. The balls were sometimes inconsistent, but they offered the best level of spin and distance for strong players. Other types of balls, made for high handicappers, emphasized distance over control and used solid rubber cores, but low-handicap golfers viewed them with distain.

Early in 2000, Nike introduced a solid-core ball aimed at tour-level golfers, which its star endorser Tiger Woods began to use. Titleist, the largest maker of golf balls, had its own solid-core model under development, which combined a large rubber core with a harder mantle layer. The outside cover was made of urethane, a soft plastic. The ball yielded the distance of solid-core balls with the control of the balata models. It was like nothing the sport had ever seen. Balata balls were very inconsistent—some seemed to fly better than others, and players would struggle to adapt to a different performance every time they'd break out a new ball. And over time, the balls would start to break down, getting out of round or cut by the club during a shot.

Solid-core balls like the Pro V1 were much more consistent and reliable. The durability was better. The solid core allowed engineers to tune the ball to react differently in different situations. When smashed with a driver, the ball would spin less than a balata ball, keeping it from hooking or slicing. When hit with a wedge, it would spin more quickly, giving the player more control to stop the ball on the green. And in every situation, it flew significantly farther than a balata ball when hit with the same force.

The first week the new Pro V1 model ball was available for tournament play, in October 2000, forty-seven players switched from their previous ball. That sort of wholesale equipment change was unprecedented in the history of golf. How fast was the transition across the sport? At the 2000 Masters, fifty-nine of the ninety-five players used a wound golf ball. One year later, only four players used one. By the end of 2001, not a single tournament champion on any of the world's major professional tours had won using a wound ball; the rout was so comprehensive that Titleist stopped making them at all.

Today, the seventh generation of the Pro V1 and its brother model, the Pro V1x, are made at Titleist's ball plant 3, in New Bedford, Massachusetts. Walking the factory floor, you're surrounded by balls in various states of manufacture, from the raw rubber to the cork-shaped billets that are then molded into spheres. There are bins and bins of centers, of balls with the covers molded on that haven't been polished, of polished and painted balls waiting to be packaged. They make three hundred thousand Pro V1s here each day, balls destined to win major titles or to find the bottom of a lake after a duffer's bad drive.

The invention of the Pro V1 started out as a little bit of an accident. The company's engineers were just trying to combine some of the technologies in their balls for amateur golfers with the ones in their pro models, and they stumbled upon the construction of the Pro V1. From that point, its refinement became a process that involved five years of prototypes and endless testing at the company's facility in Massachusetts. "We didn't have a clue what we really had at the time," recalled Bill Morgan, the company's head of golf ball development, in a 2013 interview. It took a day in which a hundred of the company's sponsored pros used the prototype ball—and gave it rave reviews—for the company to fast-track it into production.

The final leap in driving distance took place between 2002 and 2003, when PGA pros added another seven yards to their average drive. The catalyst was a jump in the size of drivers. The larger volume of the drivers allowed the face of the club to flex on impact with the ball and then spring back as the ball started to leave the face. This trampoline-like effect was pronounced in the larger clubheads, and another huge leap in distance was the result.

This was the final straw for the two organizations that control the rules of golf around the world, the U.S. Golf Association and the Royal and Ancient Golf Club, in Scotland. After years of inaction as players hit the ball farther and farther, the two governing bodies decided to cap the volume of a driver at 460 cubic centimeters, and set limits on what's called the coefficient of restitution of a club face (which is that springlike effect). And, as you can see from the graph, the massive gains in driving distance have largely stopped, with some smaller gains in the following years, and even some drops.

But the game had been completely altered in those ten years from 1993 to 2003. The average tour player gained twenty-seven yards off the tee in that time, and the longest hitters gained even more distance. The result was a totally different sport. Take, for instance, Augusta National Golf Club, perhaps America's best golf course and the annual home of the Masters. When the first Masters was played in 1934, the course measured 6,700 yards. Over the years, the club tinkered with the course, but didn't significantly lengthen it; in 2001, it was 6,985 yards long. But the driving distance of some players was starting to completely overpower the course. On holes that were designed to require a long iron second shot, players were hitting massive drives and then 8-irons to the green.

The only solution available was to lengthen the course. They stretched it to 7,270 yards in 2002, and then lengthened it again in 2006 to 7,445 yards. Today, Augusta plays at 7,435 yards. It could actually stand to be a little bit longer than that, but they've basically run out of room; instead, they have to try other methods to make the course more difficult.

"The beauty of Augusta used to be the width that was available to the golfer," says Geoff Shackelford, an author, golf historian, and course architect. "It used to be the essence of a risk-reward course, where you could be aggressive at times, and were rewarded if you could hit the shots. But they've narrowed lots of areas and really lost some of that character."

The club has used other techniques to combat distance gains. Augusta used to be known for having very firm fairways that were mown very closely—"They almost looked like greens when I first came to Augusta in 1986," recalls Shackelford. Today, the club doesn't crop the fairways as tightly, and more significantly, it mows the fairways with the mowers driving toward the tee. That causes the grass to point at the golfers on the tee, and cuts down on the roll that they get after the ball hits the ground. "This is what they have to do to deal with modern driving distances," says Shackelford. "And they shouldn't have to." In trying to control the distance of drives, the essential nature of the course has been changed.

Augusta isn't alone in this battle. Other courses, courses where the history of golf has been written, are trying to cope with the current state of the sport. Even the very place where the game was invented, the Old Course at St. Andrews, Scotland, is undergoing renovations to try and keep pace with the distance that players can hit the ball. "What other sport gets to go to the place that it all started and still have it be a relevant test for players?" says Shackelford. "That's a special thing, and we're in danger of losing it."

Now, longer drives (and longer shots with all other clubs) aren't bad in and of themselves. For recreational golfers like me, the ability to hit a drive more than 250 yards is pretty darn great, and makes a very difficult game easier. But those benefits that weekend golfers get from the equipment also raise the possibility of really distorting things at the top level of the game.

In some ways, this is a question about what we're hoping our sports accomplish. If we'd like golf to be a competition between the very smart people who run the equipment companies, we should remove the restrictions around the clubs and balls. "The equipment is, no matter what, limited by physics," says Benoit Vincent, the chief technical officer at TaylorMade, the largest golf equipment company in the world. "It regulates itself. Why not make things the best you can?"

But if we're hoping to test the skill of players, and to keep historic courses as challenging as possible for PGA Tour players, we need to rein in the technology. The simplest answer seems to be to change the one piece of equipment that's used on every shot: the ball. If you slowed down the golf ball, you could roll back some of the distance gains that have changed the game.

"Every other sport controls their ball," says Shackelford. "Tennis has actually slowed down the ball over the past few years, and the game has never been better." Prior to this switch, many players were simply dominating with their serves, leading to matches with fewer and fewer sustained rallies. With a slower ball, it was less of a serving contest and more about those long points.

One solution in golf would be to require professionals to play a ball that's had some of the distance engineered out of it, while still allowing recreational players to use longer-flying balls. The term of art for this in golf is *bifurcation*, and the USGA and R&A are opposed to the idea. They point to the long tradition of all golfers, professional and duffer alike, following the same rules. "We like that relatability that comes from playing the same sport in some of the same ways that the pros do," says Shackelford. "But that relationship is already broken with the distance they hit it."

Golf finds itself in an interesting technological trap. In the majority of sports, the most advanced equipment is only really of benefit to the elite performers. Think back to Formula 1. Just about every normal driver on the planet would be unable to complete even a single lap in an F1 car. You can't drive slowly, nursing it around the track—speed is necessary to ensure there is enough heat in the tires so they grip properly and enough heat in the brakes for them to function as designed.

Contrast that with golfers. While pros do take advantage of tech in their drivers, metal woods, and (of course) the ball, there's a real argument that golf technology does more to help the worst player than the best player. Look, Phil Mickelson could beat me playing with just a rusty 6-iron and a scuffed range ball. I'm grateful for a lower-spinning driver with a larger sweet spot that helps me keep things in play.

But the fundamental constraint of the courses at the top level seems to argue that we need to rein things in. Intellectually, I like that I play with the same gear that pros do, but it's more important to me that places like Augusta and St. Andrews and Pebble Beach remain relevant tests for the best players in the game.

The Secret Squirrel Club

One thing to know about the kind of gear that's often used by the world's best athletes: It's insanely complicated, and insanely expensive. Take, for instance, the equipment used by British Cycling at the 2012 Olympics. The bikes are glossy black carbon fiber, with no manufacturer's markings or decals, a stealth bomber on two wheels. They're not flashy, but they're very fast.

The British aren't the first country to look to custom bikes to gain a competitive edge in aerodynamics and weight. For the 1984 Olympics, Chester Kyle designed "funny bikes" for the United States that put riders in a much more streamlined position than standard bikes. In the mid-'90s, scientists at the Australian Institute of Sport developed its Superbike with a radical monocoque construction.

In reaction to these multimillion-dollar projects, the Union Cycliste Internationale, cycling's international governing body, now requires that any bike ridden in competition be available for purchase by regular riders. So to comply with these rules, there's a page at the UK Sport website that details all of the bespoke gear used by the British team, which you can buy.

There are the track frames. The description notes, "It was possible to make the Track Frame strong enough for the biggest of sprinters and use the same model for the lightest of riders, in every event from sprint to pursuit, making it very versatile."

Then there are the sprint handlebars, which have a superdistinctive design. The tops of the bars are flattened for aerodynamics, which has become pretty standard. But the curve of the drop bars is also flattened, leading to an almost organic shape that came to be nicknamed "the cobra" by riders.

These machines are the product of what's known as the Secret Squirrel Club within British Cycling, a group of engineers and designers who worked under the direction of Chris Boardman—the same man who put British cycling back on the map with his gold medal in 1992 (he left British Cycling after the 2012 Olympics). Boardman actually sells bikes marketed under his own brand name, but the bikes on the UCI's website that details approved equipment are listed as a product of Metron Advanced Equipment Limited.

Metron is the brainchild of a former Greek track cyclist named Dimitris Katsanis. After an international career as a bike sprinter, Katsanis enrolled as an engineering student at Plymouth University, studying composite materials such as carbon fiber and using that knowledge to start building frames. In December 2001, British Cycling came to Katsanis with a question: Could he build the best bike in the world for them? The first version, the MK1, was created in 2002, and went straight into use by the British team. A second frame, the MK2, was finished in time for the Beijing Olympics—two hundred grams lighter and 12 percent stiffer than its predecessor.

In all, Katsanis's bikes have been ridden to fifty-eight Olympic or World Championship gold medals. But the bikes aren't the only product of the Secret Squirrel Club. There's also the helmet that British riders use, which is listed on the UK Sport site as the Sprint helmet. According to the site, it "uses new materials to keep weight to a minimum . . . and is designed to work well even over a variety of head angles."

That bit about head angles is a nod at one of the usual problems with aerodynamic helmets that have a long tapered tail: They're great when the rider has his head in the right position, with the tail of the helmet lying between his shoulder blades. But if you turn your head or lower your chin, the helmet can actually cause drag as its tail hits the wind. The British helmet was built with an abbreviated tail section, which presumably allows for better performance as a rider moves his head.

Additionally, the helmets use a unique set of materials. Instead of the dense foam that's used for impact protection in most bike helmets, the Sprint model uses aluminum honeycomb material. It still compresses if the rider falls and hits his head, but it is lighter than the equivalent amount of foam. Crux Product Design, the firm that worked with UK Sport to design the helmets, took 3-D scans of each rider's head to ensure that each custom helmet not only fit perfectly, but was as small as possible to reduce aerodynamic drag.

The Secret Squirrel Club's reputation is so lofty that during the Olympics, the French team started speculating aloud about the "magic" wheels that the British riders used, wondering why the Brits hid their wheels so quickly after a race. Former British Cycling performance director Dave Brailsford was asked about it by the French sports newspaper *L'Équipe*, and deadpanned his answer, saying that the wheels were especially round. The joke lost something in translation, leading to headlines in the paper.

Of course, they aren't any rounder than anyone else's. In fact, they're made by Mavic, a French company. They're the same brand of wheels the French team rode. But the reputation of British Cycling to the rest of the world was such that it seemed possible they had found a way to make a more perfect circle.

To find out what all the British Cycling goodies would cost, I fired off an inquiry to them asking for a price quote—remember, according to UCI rules, this equipment has to be made available for purchase. After sending three e-mails asking for information and waiting months for a response, I finally got a handsome electronic brochure that set out the prices for this bespoke gear. The MK2 track frame costs £25,298.83; the sprint handlebars, another £23,247.88; and the sprint helmet, a cool £9,528.50. All told, that's £58,075.21, or roughly $95,000. Shipping costs, I'm sad to say, are not included.

These prices are almost certainly inflated to discourage others from buying the British gear. But the science, research, and engineering that goes into equipping these athletes means that the best gear in the world certainly doesn't come cheap, markups or not.

Poorly Suited?

Before the start of the 2014 Sochi Winter Olympics, apparel and equipment companies engaged in what's now become an Olympic tradition all its own: working to secure breathless stories about new pieces of gear designed to help athletes get a gold medal. BMW touted its work on a new two-man bobsled built to carry U.S. athletes to their first medal in the event in sixty-two years. (It worked. Steven Holcomb piloted the sled to a bronze in the men's event, while the sleds driven by Elana Meyers and Jamie Greubel finished second and third in the women's event.) Dow Chemical pointed to its help developing new sleds for USA Luge, which aimed to keep the sled's runners in better contact with the ice for more control and speed (American Erin Hamlin won the first-ever U.S. singles medal in luge at Sochi).

But the biggest tech story heading into the games was the new speedskating suit that Under Armour, the Baltimore-based apparel company, had created for US Speedskating. *Sports Illustrated*, NPR, Gizmodo—all of them wrote stories about the creation of the suit. Under Armour worked with Lockheed Martin to develop what was called the Mach 39 speedskating skin. The companies did more than three hundred hours of wind-tunnel testing on the suit, which utilized five different types of fabric, matching various textiles with different functions like reducing friction and increasing ventilation. The most radical development was a series of rubberized dots on certain areas of the suit. The dots were designed to work much like dimples on a golf ball, which help reduce the drag of airflow over a surface. It was a big investment of time, energy, and money, but the company was confident it would pay off.

"What is surprising to people is that by putting that much horsepower in we were able to improve on something there wasn't much improvement left to gain," Kevin Haley, Under Armour's senior vice president of innovation, told SI.com.

Heading into the Olympic Games, American skaters Shani Davis, Heather Richardson, and Brittany Bowe were at the top of the season-long World Cup standings. With the added advantage anticipated from the new suit, they were hoping for a very successful competition.

That's not how it turned out. Davis, the two-time defending champion in the 1,000 meter race, started with a very quick first lap, but faded badly, finishing eighth. The next day, Richardson and Bowe finished seventh and eighth in the women's 1,000 meter race (after having finished no worse than third in any of the pre-Olympics races). By then, rumblings had started not just in the media, but among the skaters themselves. Perhaps the vaunted suits, which hadn't been worn in competition until Sochi, weren't as fast as promised. Maybe the suits weren't faster than the competition's suits; maybe they were actually *slower*.

The *Wall Street Journal* brought the story to light after the disappointing results in the women's 1,000. The questions around the suit seemed to focus on a venting panel on the back of the garment that was designed to keep the skaters cool; some people thought that the panel was actually increasing drag by letting air into the suit, making it act like a parachute. Richardson sent her suit to an Under Armour seamstress and had it altered so the vent was covered by a piece of rubber, but that didn't seem to have any effect.

Other questions were raised. Should the U.S. team have trained at sea level rather than at altitude before the Sochi games, which were held at sea level? (This is actually a pretty common practice.) How did the rest of the world, including a dominant Dutch team, take such a step forward from their pre-Olympics form? Did the American team just not perform well, or peak too early in the season?

Those are all good questions, but the controversy surrounding the suits came to dominate the discussion. Eventually, the American skaters got permission to go back to an earlier suit, also made by Under Armour, which they had been wearing during the previous World Cup races. That was probably the best decision they could have made, because as soon as the skaters wondered if the new suit was slower, it was. And that has nothing to do with the technology in the suit. It has to do with the effect of belief in athletes.

Belief effects (or the placebo effect) are a powerful force in all our lives. For athletes, they can be a crucial part of being as ready as possible to compete. "Knowing you're on the start line and absolutely believing that what you're wearing and using and doing is the best in the world creates a psychological response that gives you an advantage," says Scott Drawer, the former UK Sport research chief. "The psychology is massively important, almost as important as the bit of kit itself."

That's to say that if you really think that your gear—say, a new speedskating suit—will make you faster, it probably will. As a 2013 editorial on the role of the placebo effect in sports states, "Sport scientists have often observed that just believing in a novel and exciting performance-enhancing treatment can produce improvements in performance regardless of introducing a real treatment effect."

But the flip side of that is that as soon as you start to worry that something is slowing you down, it's probably going to do just that. That's why it was a good call for the U.S. skaters to switch back to the old suits when doubts were raised about the new ones.

Unfortunately, changing suits didn't change their fortunes. The team failed to win a single medal in long track speedskating—the first time that the United States had been shut out of the medals in thirty years. No American skater finished higher than seventh in an individual event.

Near the end of the Olympics, US Speedskating renewed its partnership with Under Armour for eight more years, even as questions still swirled about the suit. As of this writing, skating officials, as well as the U.S. Olympic Committee, are digging into what happened to the American skaters in Sochi, but this much is certain: Even if the Under Armour suit is, in fact, faster than other suits, the company and skating officials are going to have to convince their athletes of that fact if they're ever going to reap any benefits from it.

A New Kind of Arms Race

There aren't just the direct consequences of these new sports technologies, like larger clubheads or straighter skis. There are less immediately obvious effects on how athletes play and train for the sports. It becomes a feedback loop—new equipment changes the sport, and then the athletes change based on the equipment, which changes the sport even more.

In golf, the larger clubheads and hotter ball didn't just alter the distance that players could hit the ball; it altered the entire approach to *how* they hit it. Players don't need to be as precise as they used to be to hit the sweet spot. In the past, golfers would often not swing as hard as they might have, just to ensure that they hit the ball squarely.

But those concerns about precision faded as clubheads got larger. Now, there is a generation of golfers who have never had to throttle down their swing with the driver to ensure good contact. That's led to a much longer swing, a wider arc that encourages a huge coil away from the ball, and then much more aggressive contact. "Grip it and rip it" has become the mantra on the driving range.

For you and me, it seems like the increased performance of something like the Pro V1 ball and bigger drivers should be a huge benefit. After all, there's no way that I could hit the ball as far as I do if I were using a persimmon driver and a wound golf ball. But contrary to what you might expect, the average handicap in golf hasn't really changed all that much as all of this gear has changed. The average handicap of a golfer was about 16.3 in 1991; by 2012 that had fallen to only 14.3. The pros might have gained thirty yards of distance over that time, but most of us aren't much better than we were before.

Why is this? Well, courses have gotten harder for all of us. They've become longer; the greens have become faster. Amateur golfers tend to play from the back tees, as most of us assume that we hit the ball much farther than we actually do. And, of course, there's more to golf than distance. For a pro who consistently hits the ball where he wants to, distance can be a huge advantage. For me, more distance might just mean that I hit my bad shots even farther off line.

Even more than the changes in technique, the equipment has flattened out some of the differences between players. The best golfers in the world—people like Tiger Woods and Phil Mickelson—wouldn't actually mind the equipment being a little harder to hit. It would create an advantage for them, because they're so talented that they could handle the increased demands on their game. (It's perhaps not a coincidence that Tiger's most dominant period took place before the explosion in distance took full effect.) Lower-tier players on the PGA Tour are the ones who benefit most from the way the equipment narrows the gap between the best and worst players. When a particular club makes it easier to hit well, the Tigers and Phils of the world aren't as hard to catch.

"What's really changed is the ability to shape shots," says Geoff Shackelford. "And the misses are much, much better with today's equipment. It's more consistent. Things like the horrible snap hook, you don't really see that at the top level of the game anymore." I wish I could say the same about my game.

Meanwhile, Ted Ligety has had to change his training to adapt to the demands of the new GS skis. "You have to apply force to the ski longer through a turn now," he says. "So it made the sport even more taxing than it was before. You need to muscle it longer."

That's meant big shifts in his work off the snow. Previously, turning the ski was an explosive movement—the skier put the ski on edge, and then applied a ton of force straight down on the ski to initiate the turn. Consequently, the training that Ligety did in the gym was focused on those requirements, meaning lifting high weights for low repetitions, focusing on building the strength to get the most out of his turns on the slope. But now you have to apply force over a longer period of time. That means lifting more repetitions at a lower weight and doing circuit training, in which he goes from exercise to exercise quickly, helping build his muscles to fight fatigue during sustained effort, rather than maximizing the force he can produce. (The same applies to all of us in the gym. Doing fewer repetitions with high weight focuses on pure strength, while lower weight for more repetitions shifts the emphasis to muscular endurance.)

By the time he arrived at the Sochi Olympics, Ligety was the huge favorite in the giant slalom. Alpine skiing is one of the most unpredictable sports in major championships—anything from putting the wrong wax on your skis to one small mistake on the slope can doom a racer to four years of frustration and questions. Ligety had been a favorite in Vancouver, before the ski rules changed, but he skied a cautious, tentative race in the giant slalom and finished ninth.

In Sochi, he skied in his typical style. Compared with the rest of the field, Ligety started his turns earlier and arced them longer, meshing his technique perfectly with the equipment that he dislikes so much. In the first of the two runs, Ligety destroyed the rest of the field, pulling out a lead of nearly one second over all the other racers. That enabled him to ski a little more carefully in the second run, as the course started to get rutted and bumpy due to the warm conditions in Sochi. Ligety ended up taking the gold medal by 0.48 seconds, and felt the weight of the world lift from his shoulders. "I've been wanting to win this medal for my whole life," he said at his press conference after the race. "So it's awesome to be able to come here and compete and finally do it and get the monkey off my back."

Even though the change in skis has been a huge boost for Ligety's competitive results, he continues to argue that it's not the right thing for the sport. "For years, ski companies used racing as a way to try out new ideas, kind of like Formula 1," says Ligety. "But no one in the general public is going to ski on the kind of skis we use today."

Remember that the stated reason for the ski regulations was to increase the safety of skiers. Ligety thinks the FIS should actually stay out of that debate and let the ski manufacturers try and innovate, arguing that those companies—who spend a ton of money sponsoring athletes—actually have the greatest incentive to develop a ski that's both safe and fast.

Ligety is right—recreational skiers will stick to the sort of shaped skis that are easier to turn and enjoy, which don't require the strength to muscle around turns. Those of us who don't make our living in sports think of equipment as a tool. Many of us—or should I say "I"—can become total gear nuts, chasing the next innovation that will suddenly improve our golf game or bike ride or make a run more comfortable. It's an arms race, but one that's about enjoyment.

But for pros like Ligety, or golfers on the PGA Tour, gear isn't just a tool. It's like oxygen, part of the conditions under which they live and train and compete. What the change in skis shows—just like the pole vault and golf before it—is that the tools and venues involved in sports aren't a neutral playing field, as we often think of them. They're important parts of determining the winner and loser, and, on rare occasions, the biggest part.

8

WHAT GETTING TIRED MEANS

Lactic Acid, Nobel Prizes, and the Power of the Mind

It must have been quite a sight as Professor A. V. Hill and his colleagues arrived at an eighty-five-meter-long grass running track in Manchester, England. It was the early 1920s, and Hill, a professor at Manchester University, was hoping to better understand the relationship between the speed a runner could maintain and the amount of oxygen the runner used. The experimental subject that day? The good professor himself.

Hill had already established himself as one of the leading scientists of his generation. He was born in Bristol, England, in 1886, to a poor family, but managed to excel academically and eventually won a scholarship to Trinity College, Cambridge. While there, Hill began to study how heat was produced by muscle contraction, using dissected frog muscles and sensitive temperature-measuring equipment. He proved that heat is produced by muscles not only when they contract but also when they are relaxed or recovering. His experiments showed that this heat production takes place even when there is no oxygen present, which was contrary to the prevailing theory at the time. He won a Nobel Prize in Physiology or Medicine in 1922 with German biochemist Otto Meyerhof for work stemming from this discovery.

Hill didn't just have an interest in the cellular level of muscular function. He was also curious how that function was manifested in athletic
performance. Today, we call that sort of science exercise physiology, but
back in the twenties, there wasn't even a name for it. Hill was actually a
fine athlete himself. His personal best of 4:45 for a mile is excellent (if you
don't think so, just go try and top it), and in one of his own papers, he
reported a VO_2 max of 57 ml/kg/min, which would project out to a sub-
three-hour marathon. In other words, he was one fit physiologist.

So it wasn't surprising that Hill chose himself as the subject at that
grass track. He strapped a bag-and-valve apparatus to his back, so that as
he ran the gases he exhaled could be captured and later analyzed. Hill ran
the track at a constant speed (with cues from a timekeeper to ensure he
wasn't speeding up or slowing down), and at thirty-second intervals, he
breathed into the valve. He started the trials at a moderate pace, and then
did three more sets, each at a progressively faster speed. After analyzing the
gas he had exhaled, and doing a lot of fancy slide rule work, Hill and his
team were able to plot the amount of oxygen he consumed over the course
of his runs.

As Hill's speed increased, he used more oxygen to fuel his efforts, as
one might expect. But his oxygen consumption eventually reached a plateau of sorts, where no matter how much faster he ran, his oxygen consumption remained the same. Once he reached that plateau, he simply
couldn't sustain the pace for very long. Hill pondered the implications of
this finding, and came up with the following explanation: Every runner
had a certain critical speed. When they were running below that speed,
they could get enough oxygen to fuel their efforts. But when they had to
run faster, the body wasn't able to process enough oxygen, and lactic acid
would build up, causing fatigue.

You can almost consider this the founding principle of exercise physiology. What Hill was trying to describe are the physiological factors that
limit our performance when we run, and his answer is simple. The key
factor is that we run out of the ability to process enough oxygen to power
our muscles. When that occurs, the muscles are no longer operating aero-

bically, or with oxygen. They're operating anaerobically (without oxygen). Hill was the first scientist to recognize that we can go into what he termed "oxygen debt" as we exercise:

> Were it not for the fact that the body is able thus to meet its liabilities for oxygen considerably in arrears, it would not be possible for man to make anything but the most moderate muscular effort. . . . It is obvious, however, that we must regard the muscle as capable of "going into debt" for oxygen.

It's hard to overstate what a profound insight this was, especially given the limited knowledge that Hill had of the actual cellular processes inside muscles. ATP, the molecule that fuels muscular contraction, hadn't even been discovered. Over the years, many of Hill's theories have been refined and tweaked, and the chemical and metabolic pathways have become more clearly understood. But the paradigm that he established has carried through much of the research to this day. Different scientists have focused on various limitations, from oxygen to ATP to glucose to glycogen, but the hypothesis has been largely the same: that once a limitation has been reached at the muscular level, there's a breakdown that leads to failure of the muscle and an inability to do any more work.

That probably makes sense to most of you reading this, because that's what we've all been taught. But there's a very real possibility that what we've been thinking about fatigue for nearly a century is all wrong, and that what limits our physical performance isn't our muscles, or our lungs, or our heart. Instead, it might be our heads.

Frogs and Fatigue

The first edition of Dr. George Brooks's textbook *Exercise Physiology* has a curious photograph on page 190. Two runners are charging around the banked turn of an indoor track. The runner on the outside of the turn, wearing a Villanova singlet, is trying to chase down his competitor. That competitor, a runner for Queens College, is powering through on the inside, driving off his right foot as he looks around the curve. The caption doesn't give the reader much information as to what's happening in the photograph or why it was chosen, simply labeling the photo "Some athletes in an indoor mile relay." Why is this picture here? And who are those runners?

The name of the Villanova runner is still a mystery to me. But the young man running for Queens College is Dr. Brooks, competing in the freshman mile relay at a track meet in New York's Madison Square Garden. His experiences as a competitive runner would lead him to discoveries about how our bodies work, and a revolution in our understanding of the role of lactate during exercise (one that contradicts the paradigm that Hill established in the 1920s).

"I wanted very much to be a track runner, and I tried really hard," says Brooks, who now teaches at the University of California, Berkeley. "I did OK, but my friends were on the Olympic team, and I wasn't. And so I asked my coach why, and he said, 'You have an oxygen debt; you have too much lactic acid.' I wasn't standing on the podium, so I had to go to the library and read about it. I was reading modern biochemistry and the old ideas about lactate, and there was no connection between the two."

The ideas he was reading about were those formed by Hill and Meyerhof in the 1920s. While Hill was running around the grass track in Manchester, Meyerhof was doing experiments trying to find out the processes that power muscular contraction. In one of the tests, he cut a live frog in half (ewww), put that semi-frog into a jar, and then immersed it in a solution so there was no oxygen available to the muscles. He used electricity to make the muscles contract until they became completely fatigued. Upon checking the solution around the muscles, Meyerhof found a high amount of a compound with the chemical formula of CH_3-$CH(OH)$-$COOH$— lactic acid (which derives its name from the place of its first discovery, sour milk).

Meyerhof reasoned that lactic acid must be a product of muscular contraction in the absence of oxygen, and that its buildup was the cause of muscular fatigue. This view of lactate as the agent of fatigue became virtually unquestioned scientific wisdom. Today, you can pick up almost any magazine story about fitness and read that lactic acid causes the fatigue you feel when you exercise, the muscle soreness that you feel afterward, and that flushing it from your body aids in recovery.

But it's possible that Hill and Meyerhof made the classic scientific mistake, mistaking correlation for causation. Lactic acid might be present at the point of muscle fatigue. But that doesn't mean it's the *cause* of that fatigue. Additionally, Meyerhof's experimental design forced the results by removing the context in which lactate actually functions. "There was nothing for the muscles to do except break down glycogen and form lactate to get energy," says Brooks. "And by cutting the frog in half and taking away circulation, then the system couldn't remove the lactate. It was like a one-way street."

Brooks likes to use a vivid analogy. "Think about a fire and a fire department," he says. "What role do you attribute the fire department? Are they the cause of the fire, or are they there to mitigate the results of the fire? There's a correlate between the firemen and the fire, and in Meyerhof's experiment there was a good correlate between the fatigue and the pileup of lactate. But they're not the same."

Brooks has spent years refining our understanding of lactate. As a postdoctoral researcher, he injected rats with sodium lactate and then made them run. Rather than seeing inflated lactate levels, Brooks found that he couldn't find any lactate in the rats' systems at all. Using radioactive tracers, he discovered that the rats were burning the lactate for energy as they ran. Lactate wasn't a waste product; it was the product of one chemical reaction that became fuel for another. "Lactate is really at the crossroads of metabolism," says Brooks.

Here's our best understanding of lactate now. We know it is produced not just by our muscles' anaerobic (non–oxygen burning) energy systems, but by our aerobic energy systems as well. In fact, these two energy systems are more tightly linked than we've previously thought; it's not an either-or system, but a matter of degree. At a lower endurance pace, we primarily (but not solely) use our aerobic system. At faster running speeds, we use more and more of the anaerobic system, while still using the aerobic system.

But both systems generate lactate. Our cells produce more and more lactate as we work harder and demand more energy output from them. Some of that lactate is shuttled, to use Brooks's term, to other parts of the body for fuel. According to Brooks, during hard exercise most of the fuel for our heart is actually lactate. "So you're gassing your heart from your legs," he says. Lactate is also shuttled to the liver and kidneys, where it's converted back into glucose (which can then power more exercise).

It isn't bad, per se, to produce a lot of lactate. However, the trick is being able to clear and process it quickly. As you train at more intense levels of effort, you are going to generate more lactate. The best athletes are those who can use that lactate most efficiently. Thankfully, that's something you can improve through training.

In the old theory of lactate as the cause of fatigue, the idea was to keep the amount of lactate produced by the body to a minimum so the athlete wouldn't get tired. But given what Brooks has shown about the body's ability to use lactate, the priority should be to increase the ability to use it during intense efforts, allowing the athlete to go faster and harder for longer.

The way to do this is with a two-pronged training approach. Lower-intensity endurance training helps create more mitochondria inside your muscle cells, which use lactate for energy. High-intensity interval training stresses the systems that transport and clear lactate. Mitochondria mass is further increased, and your body adapts like it does to any overload by increasing its capacity to process lactate.

Most recreational athletes get the intensity of their workout completely wrong. We'll go out for a run or a bike ride at a tempo that feels . . . kinda hard. Certainly not easy, but not awful either. Unfortunately, that's probably the least beneficial workout for most of us—it would be better to either go really slow, building an aerobic base, or go really, really hard, working the anaerobic systems that provide our high-end power and speed. That middle-ground workout is obviously better than nothing, but it doesn't maximize the value of the effort on either your fitness or your lactate system.

The ability to both produce and reuse more lactate is part of what helps separate great athletes from good athletes. In a study that compared trained and untrained cyclists, Brooks found that the trained riders could produce and use about 60 percent more lactate. That means that they could perform at a much higher percentage of their maximum effort for longer, generating more power and going faster than the other riders, until they reached what we have commonly called the "lactate threshold," in which the levels of blood lactate begin to spike. Reaching the lactate threshold doesn't mean you're going into oxygen debt, as many of us have been taught. At the threshold, the level of lactate is about the same as just below the threshold. It isn't that you reach a point where you start producing much more lactate than before; it's that your ability to clear and use that lactate suddenly can't keep up with the production.

Just because our understanding of lactate has been flawed doesn't mean that it's not useful. Testing for lactate threshold can still help athletes set the most effective training intensities. But what Brooks has shown is that lactate is not the cause of fatigue that it was long thought to be. So what is?

The Sky Is Pink with Purple Polka Dots

The annual meeting of the American College of Sports Medicine is the largest gathering of sports scientists in the world, drawing researchers from around the globe. Like many academic gatherings, it's a chance to renew friendships and share information, to present research and argue over its implications.

One of the highlights of each year's ACSM meeting is the opening keynote lecture, named in honor of the organization's first president, J. B. Wolffe. An invitation to deliver the Wolffe lecture is one of the highest honors in exercise science, and the list of researchers who have delivered the talk reads like a who's who of the field.

At the start of the 1996 ACSM meeting in Cincinnati, the honor of presenting the Wolffe lecture fell to a South African scientist named Tim Noakes. As an undergraduate, he had been a rower, but one day, Noakes went for a run when his rowing practice was canceled. After forty-five minutes, he experienced the "runner's high," a feeling he describes as "experiencing heaven on the road." That experience turned Noakes into a lifelong runner, and steered his professional career away from medicine and toward exercise science.

Noakes wrote the popular book *Lore of Running*, which collected a trove of scientific information for runners. He also did early research on what's called hyponatremia, the potentially fatal condition that can occur when an athlete drinks too much water during a workout or race, leading to an imbalance in the body's sodium content.

Noakes knew that delivering the Wolffe lecture was a big moment for him. Before flying to the States from London, he had taken a rare three days off to prepare. There would be five thousand or so people in the audience, and Noakes was going to challenge their foundational beliefs about what limits our performance.

He had started forming a new theory about fatigue from the time he started his own lab in South Africa in 1981. They didn't have expensive testing equipment, and still relied on chart recorders and graphs to calculate things like VO_2 max. When they would test athletes, they couldn't seem to find the plateau in oxygen consumption that Hill had talked about. "I thought it must have been a function of the equipment at first," recalls Noakes. "But then we had all of these rats that we would run and run on their wheels, and we couldn't find a plateau for them either."

At the beginning of that 1996 lecture, Noakes criticized what he thought were outmoded theories that he described as "ugly and creaking edifices." Foremost among those edifices was the work of A. V. Hill. Noakes walked the audience through a thorough examination of Hill's work, and then looked to drive home what he saw as the folly of most exercise scientists' acceptance of Hill's conclusions that the ability to get enough oxygen is what limits our muscles:

Hill's original conclusions were not supported by his own findings. Therefore, the original basis for this physiological model is without substance. If the basis for the model is in doubt, then it behooves us to question vigorously the further predictions of that original model.

This is the sports science equivalent of stepping onto a stage and proclaiming that the sky isn't blue, but rather pink, with purple polka dots. If Noakes was right that the model Hill had pioneered in the 1920s really was incorrect, then the entire field of exercise physiology had been going down the wrong path for most of its existence.

Noakes had feared that the audience would react poorly to his attack on the foundations of their field. But in the moment, he recalls, "The reaction was very positive. I thought it had gone really well, and that things had been well received." He then chuckles at the memory, knowing that the reception in the room wasn't the same as what was to follow. Criticism started to build. Noakes's lecture was published in the ACSM's journal, and in the same issue a rebuttal to the points he made was also published— "The first time there had ever been a rebuttal to the Wolffe lecture," says Noakes.

Scientists lined up on either side of the debate, defending or attacking Hill's view of oxygen's central role in fatigue. For Noakes, the easy part had been to point out the apparent deficiencies in the old model. But if the Hill model truly was flawed, the next step was to put forward an alternative model to explain what really happens as we tire.

Are Our Tanks Ever Really Empty?

It's a beautiful cloudless morning as you swing your leg over the top tube of your bicycle and head out for a ride. After a long week of work, you're really looking forward to just having a little time out by yourself, pushing your body and reveling in the physical nature of a challenge.

The start of the ride is a nice flat spin, but soon you find yourself on a long climb out of town. Your heart rate soars, your breathing speeds up, and halfway up the hill, your legs start to burn as you stomp on the pedals. You begin to play games with yourself. "Ride hard to that sign up there, and then take it easy," you think. "Now, just to that next sign."

Soon, you crest the hill and start down the other side. In minutes you're feeling great, flying quickly down the road. The miles tick by. The sun and wind in your face invigorate you, and you push the pace.

But eventually, you start to tire. You haven't ridden a lot recently, and your body is screaming. The wind shifts, turning into a headwind and leaving you struggling to maintain even what feels like a slow pace. Your world seems to shrink as you slog along, mile after mile, feeling like you're in survival mode. It's all you can do not to stop.

But as you grind back into town, just two miles from home, something changes. You feel . . . better. Good, even. You speed up, the tightness in your chest gone, your legs suddenly feeling light and powerful. The last minutes of the ride fly by, and afterward, as you look at your training data, you find that those last two miles of the ride were by far the fastest you rode all day.

Exercise physiologists have a name for that sudden burst of speed and energy at the end of a bout of exertion—they call it the "end spurt," and just about anyone who's ever exercised has felt it. There's something about knowing that we're just about to be done exercising that seems to make it much easier to push through the fatigue we might be feeling and finish the effort.

But how is this possible? If the classic model of fatigue is correct, there shouldn't be any extra muscular power left to access—after all, the peripheral model assumes that once I've depleted a muscle, it's done until it can recover.

Most of us think of an athlete's body like a car. At the start of a race, there's plenty of gas available, and over the course of the race, the athlete's effort depletes that fuel until she crosses the line. Ideally, she will have totally exhausted her resources just at the end of the race—sports writers and commentators like to refer to this as "leaving nothing in the tank."

The end spurt confounds this view. So does a study of the pacing used by athletes in world record performances. One might anticipate that the most effective strategy in a distance-running race like the 5,000 or 10,000 meters would dole out an athlete's effort evenly throughout, so that each kilometer was completed in roughly the same time.

But that's not how history is made. In sixty-six world-record-setting performances, the pacing was remarkably consistent. In sixty-five of those races, the two fastest kilometers for each runner were the first kilometer, when they were presumably at their freshest, and the final kilometer. That's the end spurt at the very top level of the athletic world.

Clearly, athletes aren't just emptying a gas tank and hoping that they don't peter out before the line. (Some research argues that athletes can refill the tank by recovering during races.) To torture the automotive metaphor further, there's something that seems to cause us to take our foot off the gas in the middle of an effort, and then to floor it when we're close to the finish.

The classical model of fatigue that stems from A. V. Hill's work is based on the idea of failure, that there comes a point when some fundamental physiological system can no longer keep up with the demands of the physical effort. Hill thought the limiting factor was oxygen, and some later researchers argued that the limit was fuel or that there were too many waste products in the muscle. But all of them have relied on the same idea that there's something at the muscular level that reaches a point where the muscle can't do any more work, and the athlete has to reduce his or her effort, or stop altogether.

According to Noakes and other researchers who proposed the so-called "central governor theory," the key factor isn't in the muscles. They argue that activity is controlled by the brain, which has just one goal: to make sure that nothing in our body is pushed beyond the normal range.

So say you find yourself at the starting line of a 10K running race on a cool morning. Based on its knowledge of your physiological ability, environmental conditions (like the weather), how long it thinks you'll be running, and your previous experience, your brain subconsciously establishes a pacing strategy that will allow you to get to the finish line without any major breakdown. From the moment you take your first stride, your brain has already decided how fast it's going to allow you to run throughout the race.

Now, if you were lining up to run a marathon on a hot day, your brain wouldn't allow you to run as fast as you would the 10K. Factoring in the heat of the day and the anticipated distance, it would select a different, slower pace—all in the service of getting you to the end of the race in one piece. The brain controls this pace by varying the amount of muscle it recruits as you run.

So if that's true, then what is fatigue? To Noakes and his colleagues, fatigue is an emotion, a construct in the mind that helps ensure that exercise is performed within the body's ability. That emotion is affected by many factors, such as motivation, anger, fear, memories of past performance, self-belief, and what the body is telling the brain. Noakes writes:

> We propose that fatigue is a combination of the brain reading various physiological, subconscious and conscious signals and using these to pace the muscles in order to ensure that the body does not burn out before the finish line is reached. I am not saying that what takes place physiologically in the muscles is irrelevant. What I am saying is that what takes place in the muscles is not what causes fatigue. Instead, metabolic and other changes in the muscles provide part of the information that the brain needs to be able to calculate the appropriate pace for events of different distance and in different environmental conditions.

In a sentence, the central governor theory claims that our physical performance is *regulated* by the brain, not *limited* by our hearts, lungs, or muscles.

Fool on the Treadmill

It's nearly impossible to "prove" anything when it comes to the central governor theory. There's no way to gather the sort of information and data that would lead to an indisputable conclusion as to whether the theory is correct or not. But there are some things that we see in labs and in the field that strongly suggest that a great deal of our performance is centrally regulated. Take, for instance, the strange case of rinsing your mouth out with carbs.

Carbohydrates are obviously a hugely important part of an athlete's fueling strategy. In study after study, athletes perform more effectively in endurance activities if they are eating carbs during them. What's more interesting is that you don't even need to actually *consume* them to get some performance gains.

A group of researchers in the UK first published this surprising finding in 2004. In the study, a group of cyclists were asked to complete an hour-long time trial while rinsing their mouths out with fluid. On some trials, they were rinsing their mouths out with water, and on others, they were using a solution that contained carbohydrates. After swishing the liquid around their mouths like mouthwash for five seconds, they spit it out, so they didn't swallow any of the carbohydrates.

The results were striking. When the cyclists rinsed their mouths with the carb solution, they performed 2.9 percent better than when they used water.

What could possibly account for this result? No carbs were entering the cyclists' systems, so it couldn't be a result of having more fuel to burn. Other studies showed similar results, and by 2009, researchers had found using MRI scans that carb rinses were activating parts of the brain associated with rewards and pleasure. (The working theory is that there are sensors in our mouths that can detect carbs in the liquid, and they tell the brain that carbs are on the way. The researchers also found that cyclists using the carb rinses had a different sense of how hard they were working, what researchers refer to as a "rating of perceived exertion" (RPE):

> The fact that the subjects in the glucose and maltodextrin [car-bohydrate] trials were working at a higher power yet reporting the same RPE as in the placebo trials, suggests that oral exposure to carbohydrate evokes a central response that enables subjects to increase their power output by reducing the perception of a given workload.

The riders were working harder when they used the carb rinse, but they didn't feel like they were working any harder. That's a pretty strong suggestion that there's something beyond our muscles that regulates our work rate and fatigue. This effect seems to work for elite athletes and normals alike; give it a try and see. Swishing some sugar water around your mouth can give you an energy boost.

If just rinsing our mouths with carbs can change our brains' perception of effort and allow us to work harder, is there other evidence that our perception can be manipulated? After all, a key component of the central governor theory is the idea that our perception of our effort is a crucial factor in our regulation of that effort.

As smart as our brains can be, we're pretty easy to trick. For instance, a 2012 study asked a group of trained cyclists to perform a four-thousand-meter time trial in the lab. The only instruction they were given was to go as fast as they possibly could. That performance was used to establish a baseline. The cyclists returned to the lab twice, but on those occasions, they raced against a virtual avatar.

In one of those races, the virtual rider's time represented the cyclist's time in the first trial. With a "competitor" to ride against, the cyclists were able to go faster than their previous effort (which you'll remember was supposed to be all-out) by about 1 percent. A small change, but instructive. But even more instructive was what happened when the researchers sped up the avatar, so it was faster than the cyclist's actual original speed. Again, the riders were able to beat the virtual racer, now 1.7 percent faster than the original. Through deceptive feedback, the riders were able to increase their performance via nothing more than their perception of how fast they could go. This sort of deception raises all sorts of interesting possibilities (not to mention ethical questions). Could I run faster if I had a stopwatch that timed each minute as fifty-eight seconds? After all, I think I can run a mile in six minutes, but maybe I could run it in 5:50 with a doctored watch.

The power of deception can also be seen when it comes to temperature. An endurance athlete's performance decreases as it gets hotter; the central governor theory hypothesizes that the brain wants to make sure the body doesn't get dangerously overheated. In a 2011 study, researchers used a rigged thermometer to see if those decreases in performance could be manipulated by faulty information. It turns out that they could: While the tested athletes' performance dropped by about 4 percent when they went from normal to hot and humid conditions in the lab, telling them that the temperature was cooler than it actually was allowed the athletes to overcome the effects of the heat. The athletes who were told it was cooler had a lower perceived effort and were subsequently able to ride harder, creating what the researchers described as "a subtle mismatch between the subconscious expectation and conscious perception of the task demands." The body has temperature controls, but the mind seems to be able to override them to some extent.

Drugs can also interfere with our perceptions of temperature and effort. British cycling champion Tom Simpson famously collapsed and died as he was climbing Mont Ventoux in the 1967 Tour de France. On a day when the temperature was above 100 degrees, Simpson had taken amphetamines before starting the climb of the mountain. Near the top of the climb, he began to wobble and then fell off his bike. His team's mechanic tried to get Simpson to abandon the race, but he insisted on continuing. He remounted, carried on a few hundred more meters, and then toppled and was caught by three spectators, who helped him to the ground by the side of the road, with his hands still locked on the handlebars. He wasn't breathing.

Simpson was airlifted off the mountain to a hospital in Avignon, but it was too late. He was dead at age twenty-nine. In the back pocket of his jersey, there were three identical tubes, two of which were empty. The third was half-full of amphetamines. Simpson's official cause of death was heatstroke. But the culprit was likely those amphetamines.

At that time, use of the drug was common in cycling, but the outcome for Simpson was tragic. Lab tests have shown that amphetamines can allow athletes to exercise at higher intensities and for longer durations than normal, and more meaningfully, they can sustain greater levels of muscular and cardiovascular work. But the perception of that work remains the same. It appears that amphetamines short-circuit the brain's perception of effort, potentially allowing athletes to push themselves into places their bodies can't handle.

These effects might be exacerbated by extreme temperatures. Antidepressants like Wellbutrin and Ritalin have similar effects in heat—these drugs seem to disable the "safety brake" in the brain that usually stops athletes from pushing into the temperature danger zone. It's a good reminder for anyone taking these drugs to be careful of their effects in heat.

We can even be fooled by social psychology.

In a delightful experiment that pointed out just how easy it is to influence human beings, male runners were asked to do three different trials of a twenty-minute running task. In one trial, an attractive female observer entered halfway through and stood with the researcher conducting the test; in another trial, an attractive male observer did the same thing. When the runners' RPEs from the two tests—conducted at exactly the same intensity—were compared, the research team found that the male runners reported a lower perceived effort while being watched by a woman, and a higher perceived effort when being watched by a man.

The researchers write: "The finding that lower RPE scores were recorded when a female observer was present suggests that the motivation to portray physical competence (enthusiastic, alert, and free from physical distress) may have been increased compared to when an observer of the same sex was present."

This may seem quaint, even stereotypical, in its seemingly outdated notion that men want to impress women. But the results are there, and they hint at something larger. Our bodies and minds are so entwined that there's no way to separate them. And even if we'd like to think that an amount of effort is a constant, the presence of a pretty woman can seemingly change that, or at least our perception of it. And if something so trivial can have such an effect, what else can we do to fool our brains into allowing us to go longer and faster?

Tired Brain, Tired Body

Samuele Marcora is a researcher at the University of Kent, in England; previously, he was at Bangor University, in Wales, where he began to study and demonstrate the connections between mental fatigue and physical performance. His group compared the performance of cyclists who had done one of two things before starting an exercise test: They spent ninety minutes either performing a letter-recognition task or watching documentaries about Ferraris and the Orient Express train. Then, they were asked to get on a stationary bicycle and perform a standard physiological test until they couldn't go on any longer.

The riders who had watched the movies averaged about 12.6 minutes in the cycling test. Those who had done the cognitive test, which was designed to fatigue them mentally, lasted only 10.7 minutes on average. The subjects who were mentally fatigued performed about 15 percent worse than the other riders.

There wasn't any sort of difference in the heart rates or oxygen consumption between the two groups. What was different was their perception of how hard they were working. From the very start of the cycling test, the mentally fatigued group rated their effort as more taxing than the movie-watching riders.

As the researchers put it: "Given no effect of mental fatigue on potential motivation in our experiment, the key to understand its negative effect on short-term endurance performance is the higher perception of effort measured during high-intensity cycling exercise. . . . Overall, it seems that exercise performance is ultimately limited by perception of effort rather than cardiorespiratory and musculoenergetic factors."

This insight has led Marcora to propose an alternative to the central governor theory that he calls "the psychobiological model of exercise tolerance." Marcora argues that the perception of effort—RPE, if you recall—is the key limiter of our endurance, and that unlike in the central governor theory, the body doesn't have any overall system that's meant to keep it safe from harm. In some ways, this is a very real distinction when it comes to the mechanisms involved. But to most athletes, the distinction is more academic than practical. Both Marcora and Noakes put the brain at the center of the system, and they both focus on the perception of effort as a key factor.

Marcora's study suggested that the most likely area of the brain controlling this perception is the anterior cingulate cortex, which integrates motor control, emotional regulation, and cognitive function. The researchers hypothesized the ACC is involved because it's an area of the brain that's taxed by the sort of mental exercise that affected the athletes in the study—but they couldn't state definitively that it was the right area. "Neuroimaging studies are needed to confirm our hypothesis," the authors wrote.

That sort of neuroimaging study would be a chance to offer physical proof of the central governor theory. If we could come up with a way to look inside the brain of an athlete as he or she is exercising, we could look at what parts of the brain are more active than others. The fMRI machine does this by mapping blood flow in the brain, but doing actual exercise inside an fMRI machine is impossible at this time. Most studies that relate brain and physical activity have relied on small motor tasks, like making a fist.

To overcome the machine's limitations, a group of Brazilian researchers working with Noakes built a cycling ergometer—basically an exercise bike—that subjects could pedal as they were lying in the fMRI tube. It was a clever hack, built with nonmagnetic parts and an elaborate system to connect the crankset in the MRI room to a power meter outside the room, allowing the athlete's workload to be measured. Test subjects pedaled the bike at set levels of intensity based on their weight, alternating two minutes of hard pedaling with sixteen-second rest periods. Every minute, they were asked to report their RPE; later, those reported levels of effort were matched up against the fMRI activity at that moment.

The study found that during cycling, there was increased activity in the primary motor cortex, which isn't a shock, given that region of the brain is in charge of planning and executing movement. But during what the participants classified as hard exercise, the MRI scans showed that specific areas of the brain showed increased activity: the posterior cingulate cortex and the precuneus, the same general brain areas identified as possible areas of control by Marcora's mental fatigue study.

These areas of the brain are thought to be centers where higher-order cognitive tasks take place, where information from the rest of the body and brain are integrated and evaluated. What does that mean for the central governor theory? It could be that signals from the body—information about heart rate and respiration, levels of lactate concentration, the responses of muscles to their contraction—are processed in the posterior cingulate cortex and precuneus and translated in the brain as a perception of the level of effort. The authors claim this was the first study ever done that could assess the entire brain during real exercise, and the results seem to point at the very sort of brain-based regulation of effort that Noakes and Marcora postulate.

Mind Over Matter

So, what if we have been wrong about fatigue for the past hundred years, blaming it on our hearts and lungs and muscles, when it's really been about our minds and emotions? Is there some way that this new perspective becomes actionable for athletes, coaches, and weekend warriors, rather than just a different framework through which to understand performance?

The tension between the lab and the competitive arena is an ongoing one in sports science. No matter how clearly the science might indicate the effectiveness of a training method or the superiority of one framework of fatigue over another, it's essentially meaningless in competition if there's no direct way to apply it in that context.

We humans are endlessly manipulatable. The past several decades have brought dozens of insights about how our perceptions of the world are faulty, about how we can convince ourselves of things that aren't true while plainly missing things that are right in front of us. Our brains, for all their wonder, have their own shortcomings.

But athletes might be able to use those shortcomings to their advantage. If mental fatigue can lead to physical fatigue, as Marcora's study shows, then brain training—using exercises to increase the brain's "fitness" so that you won't get mentally fatigued as quickly—might be a way to improve athletic performance. If you could find a way to postpone mental fatigue, you would have a physical advantage.

Marcora has started to research this possibility with the British Ministry of Defense, which is interested in seeing if it can reduce mental fatigue in soldiers. At present, Marcora can't talk about what he's uncovered, but he and I were able to talk about what he's found out about the effects on endurance performance, which was recently declassified.

In the twelve-week study, two groups of fourteen soldiers each trained on stationary bikes. The first half trained three times a week for one hour at a moderate aerobic pace. The second half did exactly the same intensity of training for the same duration, so the physiological work was the same. But while this second group pedaled, they were also doing a mentally fatiguing task.

The results at the end of the study were mind blowing. The two groups saw similar increases in their VO_2 max, meaning the physiological effects of the training were about the same. But when you asked them to do what's called a "time to exhaustion test," in which they rode at a specific percentage of their VO_2 max until they couldn't go on, the differences were profound. The control group saw the time to exhaustion improve 42 percent from their results before the training started. The group that combined training with mental exercise saw an improvement of 115 percent, almost three times the improvement that the control group saw. Combining the physical and mental stress lead to a quantum leap in performance.

"The only thing I know that has such a big effect on endurance performance that's been published on the same kind of test is a study that Marcus Amman did with people doing the time to exhaustion test using 100 percent oxygen," says Marcora. "If you breathe 100 percent oxygen, you increase your time to exhaustion by 110 percent." Something about increasing the brain's fitness led to a massive increase in the riders' physical abilities as well.

There are even larger implications when you take these ideas about the brain and fatigue into a competitive setting. Noakes has a hypothesis about what cause athletes to win and lose. He thinks it's not physiology that ends up separating competitors, but the mind. Remember that Noakes believes that what we feel as fatigue is an illusion, one generated by our minds. Given that, he argues:

> The winning athlete is the one whose illusionary symptoms interfere the least with the actual performance—in much the same way that the most successful golfer is the one who does not consciously think when playing any shot.
>
> In contrast athletes who finish behind the winner may make the conscious decision not to win, perhaps even before the race begins. Their deceptive symptoms of "fatigue" may then be used to justify that decision. So the winner is the athlete for whom defeat is the least acceptable rationalization.
>
> How athletes and coaches achieve this winning mental attitude is the great unknown. But if the study of the purely physiological basis of fatigue has taught us anything, it is that such studies will never provide an adequate answer.

Perhaps I find this so compelling because it resonates strongly with my experiences as an athlete. Thinking back to races that I've lost, I can almost pinpoint the moment when I resigned myself to a specific outcome for the event. At the time, I viewed it as simply accepting reality—I just wasn't fast enough or strong enough to win.

But there was another dynamic involved in those moments: I wasn't willing to push through the pain and fatigue that I was experiencing. I wasn't willing to sacrifice enough. And you know what? Once I made those decisions, my sense of fatigue seemed to dissipate. I felt better, even as I knew I wouldn't emerge as the victor.

"So the winner is the athlete for whom defeat is the least acceptable rationalization," writes Noakes, and I realize that defeat was an acceptable rationalization for me in those moments. I did just what he said, consciously accepting my finishing position.

Noakes says he has talked to coaches in Canada and the UK who have designed their training programs with his insights in mind, although he won't elaborate on their methods. And Marcora has been working with the English Institute of Sport, where he's developed new training programs based on his research. "There are some things we're doing that I'm excited about but I can't tell you about it until after the Rio Olympics," he says.

In some ways, this view of the mind is something that athletes have been expressing for decades. Roger Bannister, the first man to break the four-minute mile, once said, "It's the brain, not the heart or lungs, that is the critical organ." Paavo Nurmi, the great Finnish runner from the 1920s, said, "Mind is everything. Muscles are pieces of rubber. All that I am, I am because of my mind."

Even beyond the construction of an athlete's physical training, Noakes's central governor theory points to the importance of bringing together the physiological and the psychological. "I think the key is that you have to have self-belief," says Noakes. "You have to believe that it's your right to win. You have to believe that it's your destiny. Somewhere along the way, you have to inculcate a belief that you're invincible."

This is the merger of the two worlds of the body and the brain. If fatigue is a self-generated emotion, as Noakes argues, then the greatest performance enhancer available to us isn't a drug or nutrition. It's our own minds.

9

STAYING STRONG

It's Hard Work Taking It Easy the Right Way

There aren't many places on the planet that are harder to get into than the Olympic Village during the games. Without the right credentials—and those are strictly limited to the athletes and select national team coaches—you're facing some seriously long odds to find your way into the area, which is meant to be a place for the world's elite athletic competitors to meet, mingle, and foster goodwill (that's not all they're fostering, if the reports of a hundred thousand condoms being distributed during the London games are to be believed). At most international sporting events, athletes can bring their own support crews, from coaches to trainers to doctors, but the Olympics, and the accompanying huge logistical barriers, make life a little less convenient.

The village at the London games was located in Olympic Park, in the East London neighborhood of Stratford. Also in the park were many of the games' venues, including the Olympic Stadium and a gigantic shopping mall built by the Australian developer Westfield. The idea was that visitors who arrived through some of the new transit connections to Stratford would be able to visit the mall as well as attend the events; after the Olympics, the entire area would be converted into a huge new neighborhood, complete with twenty-eight hundred housing units from the former Olympic Village.

At 5:00 A.M. on the morning of August 9, 2012, a very fit man left the Olympic Village. Ashton Eaton had already completed the first day of the decathlon competition at the Olympics, running the best 100 meters ever in an Olympic decathlon and hitting a long jump of more than eight meters. Those performances helped stake Eaton, the recently crowned world record holder in the event, to a 220-point lead heading into the second day of the competition, with five events scheduled.

But at that moment, Eaton was sore. Decathlon isn't just a battle of performance in each event, it's a battle of endurance and recovery as well. Decathletes need to make sure that they conserve as much energy as possible—in fact, Eaton's coach, Harry Marra, sometimes has to remind him to sit down between events instead of pacing or bouncing around the room. Part of Eaton's regular preparation is massage and soft tissue treatment, which he generally does twice a week. "If I didn't have that, I couldn't practice at the same level of intensity that I do," he says.

So now, before the biggest day of competition of his career, the day he might be crowned Olympic champion, Eaton wanted to get a massage. There was just one problem. His physiotherapist wasn't able to get a credential to enter the Olympic Village. That's why Eaton, as the sun rose in London, was leaving the village and heading toward the mall.

Eaton began his warm-up routine on the road to the shopping center, and then met his physiotherapist, who set up a table to do his treatments in the middle of a side street. "I was lying there thinking that this was the day I was going for a gold medal," says Eaton. "And my therapist was saying, 'I can't believe that I'm here massaging the potential gold medalist here in the middle of the street.'" Curious passersby slowed to check them out; eventually, a security guard stopped to ask what was going on. "We just waved and said, 'Nothing to worry about here,'" says Eaton. "'We're just taking care of some stuff.'"

Eaton's early morning massage shows the priority that athletes place on recovery. And it also demonstrates how much room there is to optimize how and when athletes get that sort of help.

The Queen of Recovery

Shona Halson is, as one fellow researcher calls her, the queen of recovery. Her more formal title is head of performance recovery at the Australian Institute of Sport, where she presides over two other PhDs and four other researchers at the AIS Recovery Centre. Just around the corner from her office is a large room filled with exercise bikes, stretching mats, and a private massage area to one side; it's what they call the "dry area" at the facility. Through a glass door is the "wet area," a massive room filled with various pools, showers, and tanks to give athletes every conceivable hydrotherapy option.

The Recovery Centre is a multimillion-dollar investment with a single purpose, according to Halson. "The goal is to make athletes go faster," she says. But unlike with other coaches and researchers, Halson's job isn't to improve an athlete's performance in an absolute sense. There are two ways to make an athlete more effective in competition. The first is, well, to make them better. You can train to become stronger, faster, more skilled. You can develop your tactical sense, your biomechanics, your physiology. It's what we mostly think of when we think of athletic progress, and it's related to much of what we've covered thus far in this book.

Halson and her team do something different. They are focused on making sure that the athletes in their care can access the full range of their talents and abilities when they need to. Take two athletes, both of whom can run a ten-second hundred-meter dash. Send them down the track again and again, day after day—but after the races, let one come to Halson's Recovery Centre, while the other goes home. Very quickly, the better tended athlete will completely outpace his rival. Simply put, recovery is focused on making sure that an athlete can do his or her best work, like a pit crew dealing with a high-performance race car.

For years, athletes have used some of the techniques that Halson and her team deploy, whether it's a massage for a Tour de France cyclist at the end of a hard day's racing or an NFL player lowering his massive frame into a tub filled with ice to quell the pain and inflammation in his legs after a game. But in many ways, the practical applications of recovery techniques have run ahead of the basic science. As Halson puts it, "Recovery is where carbohydrates were twenty years ago," meaning that we know a fair amount about what works, but very little about the exact mechanisms involved.

The AIS recovery team is out to change that. Working with the hundreds of athletes under their care, they're striving to understand why some athletes can work hard day after day, while others break down and become injured, or simply can't perform as they should. Using everything from that array of tubs to high-tech compression clothing, there's nothing they won't try to give their athletes an edge. But the first, and most important, tool at their disposal is nothing more than a bed.

Hit the Snooze Button

Sleep is, well, kind of amazing.

A common medical textbook breaks it down to its essence: "Sleep is a reversible behavioral state of perceptual disengagement from and unresponsiveness to the environment." It then goes on to note that sleep "is typically (but not necessarily) accompanied by postural recumbence, behavioral quiescence, closed eyes, and all the other indicators one commonly associates with sleeping." (I'm not sure why I'm so charmed by the contrast between "postural recumbence" and "all the other indicators one commonly associates with sleeping," but I can't read that sentence without laughing out loud.)

For us humans, sleep is completely crucial to proper functioning. As we've all experienced, we're simply not as adept at anything in our lives if we don't sleep well. Without proper sleep, whether it's a short-term or long-term deficit, there are substantial effects on our mood, our mental and cognitive skills, and our motor abilities. When it comes to recovery from hard physical efforts, there's simply no better treatment than sleep, and a lot of it.

Most research on the effects of sleep on athletes has studied sleep deprivation. And those effects are quite strong. Just like the rest of us, athletes see a drop in their performance across all sorts of measurements if they are kept awake for the entire night, or even just interrupted in their sleep.

It seems like certain kinds of athletic tasks are more affected by sleep deprivation. Although one-off efforts and high-intensity exercise are impacted, sustained efforts and aerobic work seem to suffer an even larger setback. Gross motor skills are relatively unaffected, while athletes in events requiring fast reaction times have a particularly hard time when they get less sleep.

But instead of focusing on the effects of a lack of sleep, it's more interesting to explore additional sleep as an advantage. If an athlete gets more sleep than his or her competitors, will that lead to an edge? That's just the question that Stanford researcher Cheri D. Mah set out to answer. She reached out to athletes at her university, trying to find a group that would participate in an experiment in which they would first measure their athletic performance after having their normal amount of sleep, and then spend weeks trying to extend their sleep as much as possible, to see what effect it would have on objective measurements of athletic performance. Amazingly, no one had ever done a study to see the effect of sleep extension on competitive athletes.

The Cardinal men's basketball team volunteered to be Mah's study cohort. Eleven players used motion-sensing wristbands to determine how long they slept on average—just over 6.5 hours a night. For two weeks, the team kept to their normal schedules, while Mah's researchers measured their performances on sprint drills, free throws, and 3-point shooting. Then, the players were told to try and sleep as much as they could for five to seven weeks, with a goal of 10 hours in bed each night. Their actual time asleep, as measured by the sensors attached to their wrists, went from an average of 6.5 hours to nearly 8.5 hours.

The results were startling. By the end of the extra-sleep period, players had improved their free throw shooting by 11.4 percent and their 3-point shooting by 13.7 percent. There was an improvement of 0.7 seconds on the 282-foot sprint drill—every single player on the team was quicker than before the study had started.

A 13 percent performance enhancement is the sort of gain that one associates with drugs or years of training—not simply making sure to get tons of sleep. Mah's research strongly suggests that most athletes would perform much better with more sleep—if they could get it. But it's not quite that easy; in fact, athletes face challenges with their sleep that many of us don't have.

The first challenge that many elite athletes face is the travel demands of their sport. When you're a pro athlete, you spend a lot of time on the road. If you're a professional sports team athlete in the U.S., you're spending your time zigzagging across the country, flying back and forth to meet the demands of schedule makers who don't always take the travelers' circadian rhythms into account.

The mileage can pile up in a hurry, especially for teams on the West Coast, which are farther away from the rest of the teams in their leagues. West Coast teams perennially have to travel more miles than their competition—in 2013, the Seattle Mariners flew more than 52,000 miles while the Chicago White Sox, with their central location and nearby division rivals, only flew about 23,000. Some years, the L.A. Kings have had to fly more than 55,000 miles to reach other teams in the NHL, while the New Jersey Devils were clocking less than 29,000. Bouncing around the country, leaving late, arriving early, having to play the next day—it's no surprise that travel and the management of sleep is a huge issue for athletes.

Imagine you're an NBA player. You play a game starting at 7:30 P.M. on the West Coast, and the superphysical battle ends at 10:00. You shower, do interviews, work on your physical recovery program, eat something, and get on the team bus to the airport. Your flight leaves after midnight, and after a couple of hours on the plane, you're back home at 3:30 or so. You've got a game the next night, so the coaches have scheduled a 9:00 A.M. walk-through; you'll be lucky to get four hours of sleep before practice.

To try and deal with this disruption, teams have consulted with sleep researchers like Mah. Most NBA players have adapted by taking a nap in the afternoon, between morning practice and the evening's game. "If you nap every game day, all those hours add up and it allows you to get through the season better," NBA all-star Steve Nash told the *New York Times*. "I want to improve at that, so by the end of the year, I feel better."

Domestic travel is bad enough, but for athletes in many Olympic sports, there's a heavy dose of international travel as well. Randy Wilber, at the U.S. Olympic Committee, notes that there's very little published research on how to deal with jet lag even for people who travel professionally, like pilots, let alone research on how to minimize its effects on elite athletes. "We've had to develop those protocols ourselves pretty much from scratch," he says.

In a guide distributed to athletes before they traveled to the 2011 World Track and Field Championships, in Daegu, South Korea—sixteen time zones away from the West Coast of the United States—the USOC scientists laid out strategies to minimize the disruptive effects of the travel.

First and foremost, the athletes were advised to stay awake on their flight to Korea. Most of the flights left in the late morning on the West Coast, and arrived in the midafternoon the following day in Korea. By staying awake, they could push through to the evening in Daegu, getting on the correct sleep schedule quickly. For these athletes, it's not just about getting acclimated to the right schedule but also about trying to get back to normal training. The guidelines suggest that they do two light workouts the first day after they arrive, and two slightly harder workouts the following day, making sure to train outdoors to maximize exposure to sun and light, which helps adjust our internal clocks. By the third day, the guidelines suggest that the athletes are ready to try harder training. The same advice applies to all of us when we get jet-lagged: Do the best you can to reset your internal clock, and get as much sun and light as possible to help you along.

Does traveling a long distance have a demonstrable negative effect on teams? There haven't been a lot of studies, but Bill Barnwell, of Grantland

.com, looked at fifteen years of data for the NFL, examining the winning percentage of road teams by the distance they had to travel:

Travel Distance, One-Way	Road Team Winning Percentage
2,000+ miles	.398
1,000-1,999 miles	.403
0-999 miles	.430

Bad news for teams like the Oakland Raiders, who had to travel more than twenty-eight thousand miles in 2012, while teams like the Indianapolis Colts only traveled 8,494 miles.

And these travel effects seem to accumulate over the course of a season. Researchers at Vanderbilt University examined the plate discipline of hitters in baseball over the course of the season, and found that hitters swing at more pitches outside the strike zone late in the season than they do earlier in the season. Why? Dr. Scott Kutscher, the leader of the research team, said in a press release, "We theorize that this decline is tied to fatigue that develops over the course of the season due to a combination of frequency of travel and paucity of days off."

Kutscher's team has found that this decay in plate discipline has become more pronounced in baseball since 2006—the year that Major League Baseball banned stimulants. (For years, bowls of amphetamines, known as "greenies," were a fixture in baseball clubhouses.) Out of the thirty teams in Major League Baseball, twenty-four saw this decrease in 2012, the year the study examined. That suggests that if a team can find a way to stem this fatigue effect, they might have a competitive advantage—in fact, it's already happened. The San Francisco Giants actually *improved* their plate discipline over the course of the 2012 season, and the team went on to win the World Series.

Another key challenge is that elite athletes seem to have poorer sleep quality than the average person. A research team at the English Institute of Sport used the same type of wrist sensors as Mah to measure the sleep of a group of forty-seven Olympic athletes, comparing it to the sleep of a control group of nonathletes. The athletes spent more time in bed than the control group, but it took them longer to fall asleep, and they had a poorer quality of sleep once they did nod off. (As a side note, the female athletes were better sleepers than the men, a gender difference that's been relatively consistent across sleep research with athletes and nonathletes alike.)

A 2011 German study found that 65 percent of athletes surveyed said they had suffered from a poor night's sleep before an important competition at some point in their careers—a percentage that seems low, if anything. They reported that the main factors that kept them from sleeping well were "thoughts about the competition" and "nervousness about the competition," which makes sense to anyone who's ever lain in bed wide awake, thinking about a problem that they need to solve the next day. That anxiety was found to be stronger in individual-sport athletes than team-sport athletes, perhaps because they bear sole responsibility for the outcome of their events.

We've talked about the disruption of sleep caused by team sports that take place in the evening. But what about those sports, like golf, running, triathlon, and cycling, that often take place early in the day? There are some good reasons why you might want to have a long-duration event start at the crack of dawn—so you have time to do anything else during the day, for one. But it also might be that people are drawn to these sports specifically because they are morning people.

There has actually been a demonstrated link between whether you're a morning or evening person and a polymorphism on the period3 gene. (Yes, whether or not you like getting up early might actually be genetic.) Researchers in South Africa tested a group of runners, cyclists, and triathletes for this polymorphism. Sure enough, more of the athletes had the genetic trait associated with being a morning person than those in the control group, and more of them identified themselves as morning people. It suggests that there's a competitive advantage for those athletes who pre-

fer being up early—and that there's a genetic component to that prefer-
ence.

So, what's an athlete—or anyone else—to do when it comes to maximizing the value of sleep? The same advice that sleep specialists give to
athletes is the advice we can all follow to sleep better without resorting to
pharmacological solutions. There are things to avoid before sleep, like alcohol and caffeine. Foods containing the amino acid tryptophan (including milk, meat, eggs, and cheese) may also help with sleep, as can a
high-carbohydrate meal four hours before bedtime. It's also good to reduce
fluid intake in the period before going to bed, as having to wake up to
urinate is a common cause of poor sleep.

It's all part of what researchers call "sleep hygiene"—making sure that
your bedtime is as regular as possible, removing the bright digital clock from
your bedside table (studies show the light disrupts sleep), finding a comfortable temperature (research shows a cool room is best). I've tried to incorporate sleep hygiene into my life, and while I haven't yet turned into the sort of
star sleeper that I'd like to be, I do feel more refreshed and better rested in the
mornings. Start viewing sleep as a performance booster rather than a chore,
and the effort it takes to sleep well will seem like a smart investment.

Squeeze Play

Beyond sleep, there is a variety of tools and techniques that you can use to
help recover from physical exertion so you're ready to work hard or compete again more quickly. One of the recovery tools that's exploded onto the
scene in the past decade has been compression garments—they're not just
for your mother's varicose veins anymore. From pro sprinters to people in
your running club, it seems like athletes everywhere have started to cram
themselves into supertight clothing, from socks to tights to shirts, all of
which promise to help you perform better when you're wearing them and
feel less fatigued afterward. But a pretty important question remains: Does
this compression gear work?

Compression garments were developed as a therapeutic treatment for conditions caused by circulation problems like pulmonary embolisms and deep vein thrombosis—the idea was that by using compressive fabrics wrapped around a body part, you could help increase circulation by using more pressure at the end of limbs and less pressure closer to the heart. The gradient pressure would help push blood back to the heart, augmenting its efforts. They were the sort of thing you would buy at a medical supply store, thick flesh-colored stockings that were nobody's idea of high-tech sports equipment.

But if these sorts of clothes could aid circulation in an unhealthy body, could they help a body at the peak of its abilities?

The first study to examine the issue was published by two University of North Carolina researchers in 1987. They found a decrease in the concentration of blood lactate of subjects wearing compression stockings during intense exercise—it appeared that the test subjects were clearing lactate more quickly when they wore them. Suddenly, it seemed like compression could be a silver bullet for athletes. The story hasn't turned out to be that simple, however, even as compression gear has become a massive industry in the sports world, helping to fuel the growth of companies like Under Armour, which launched in 1996 with a compression T-shirt that the company's founder sold out of the trunk of his car, and is now a $1.5 billion company competing with established giants like Nike and Adidas for global domination.

Part of what makes compression so hard to understand is that there are multiple hypotheses to explain its possible effects. The first is centered on the circulation-related (hemodynamic, if you're feeling fancy) effects compression could have. There are three different possibilities here. First, compression could increase the return of blood to the heart through our veins, increasing the stroke volume of blood through the heart. Second, compression could allow the flow of oxygenated blood from our arteries to increase, giving muscles more oxygen to do their work. And finally, there's some suggestion that it could increase the outflow of lymphatic fluid from muscles; this would lead to a decrease in swelling and pain after exercise. Whew.

That's not all. There are thermal effects from the garments; they keep us warmer, which (in most temperature conditions) is a good thing for our bodily and muscular function. There are neural effects: The tight fit of the garments could increase what's called proprioception, which is our sense of where our bodies and joints are in space. That could lead to better coordination and performance. There are also possible mechanical benefits. By reducing the vibration and oscillation of our muscles, compression can decrease the number of muscle fibers the body recruits for a specific action and thus lessen fatigue.

And then there are the psychological factors. Part of what's hard about compression research is separating out demonstrable physical effects from the possible placebo effects of the gear. If you ask the athletes who wear it, most will tell you compression gear just feels good to wear. It's like a full-body hug from your clothes. Even if there's no physiological effect from the clothes, the simple belief that they might help could be a powerful performance boost for an athlete.

So, given all that, what's the best evidence?

Right now, the scientific research on compression suggests the following:

1. Compression worn during exercise offers a small boost to performance on short sprints, vertical jump, and time to exhaustion in sprints and other maximal types of exercise. A basketball player wearing compression socks and calf sleeves? It seems like they're onto something real. There appears to be no significant effect on endurance sports.

2. Postexercise recovery is where compression shines. There are small to moderate effects in reducing muscle swelling and perceived muscle pain after exertion. Again, these benefits seem to be greatest after power workouts as opposed to endurance workouts. And the results are much better when compression garments are worn for a long time—from twelve to forty-eight hours after the workout.

3. Compression garments have a significant effect on body temperature—more clothing results in a warmer athlete. In a hot climate, this could be an issue, but it raises interesting possibilities for winter sport athletes. As one review puts it: "So far, no study has investigated the effect of compression clothing in winter sports. Since the reduction in skin blood flow would increase blood volume in the working muscles, compression might especially serve as an ergogenic aid in performance in cold environmental conditions."

The one other piece of good news is this: There doesn't seem to be a downside to wearing compression gear; no study to date has found any detrimental effects from its use. So, if it feels good—and your wallet can sustain a compression gear habit—go for it. After all, belief effects are a powerful thing, and if you believe that compression will increase your performance, it likely will. Just know that the science still hasn't caught up to the hype.

Vitamin I

To be an athlete is to be in pain, at least some of the time. When you push your body in training, trying to get it to adapt to more and more demands, it can be somewhere between uncomfortable and excruciating. Our bodies are simply not meant to do some of the things we do. Humans aren't designed to run at full speed into one another (covered with protective gear or not), but hundreds of men do it each weekend in the NFL season, and they pay the price through the pain they suffer.

The current thinking about pain is that our individual thresholds are relatively stable, but our tolerance—how we cope with the pain we feel—can be trained and improved. Although we can all increase that tolerance, athletes train it more than most of us do. A 2012 meta-analysis of studies on athletes and pain found that athletes consistently demonstrated a higher pain tolerance than normally active people. As the authors of the study write:

Athletes are frequently exposed to unpleasant sensory experiences during their daily physical efforts, and high physical and psychological resistances must be overcome during competitions or very exhausting activities. However, athletes are forced to develop efficient pain-coping skills because of their systematic exposure to brief periods of intense pain. Therefore, pain coping is an integral part of athletic training, and coping skills are important features in the development of athletic character. Moreover, the mental attitude of athletes towards pain and physical discomfort significantly differs from that of normally active controls.

Combat-sports athletes learn how to handle getting punched and kicked; baseball players find ways to shake off getting plunked by a ninety-five-mile-an-hour fastball; hockey players lose a tooth after getting hit by a puck and then are back on the ice for the next shift.

Another way athletes have traditionally dealt with pain is by using analgesic drugs like ibuprofen, naproxen, and aspirin. These compounds are classified broadly as NSAIDs (nonsteroidal anti-inflammatory drugs), and they all work in roughly the same way, by interfering with enzymes in the body that help create prostaglandins (which cause pain and inflammation).

Athletes *love* these drugs. A study of players in the 2002 and 2006 soccer World Cup found that more than half of them took an NSAID during the tournament. Ten percent of players overall were taking them before every match—on one squad, twenty-two of the twenty-three players were doing so. In endurance sports, ibuprofen use is so prevalent—up to half of competitors in one popular ultramarathon race took ibuprofen during the run—that it's often known as "vitamin I."

There are a couple of problems with this type of widespread use. The first is that taking ibuprofen before an event doesn't help with performance. In fact, there have been studies that have shown that cyclists perform about 4.2 percent *worse* in a ten-mile time trial when they've taken ibuprofen before the effort as compared to a placebo. Furthermore, animal studies have shown that taking ibuprofen during training can lead to a reduction in the benefits you get from it—even if you increase your training volume, you don't get the same results as you would without the ibuprofen. Ibuprofen seems to, paradoxically, increase the amount of inflammation seen in the body during exercise. And then there are the problems that chronic ibuprofen use can cause with the liver and gastrointestinal system.

Simply put, most athletes, both professional and recreational, are using ibuprofen and other NSAIDs incorrectly. NSAIDs make sense for acute injuries, in which they can help an athlete manage the pain and recover from injury. But using these drugs regularly in hopes of reducing pain before it starts isn't a good idea. As an editorial in the *British Journal of Sports Medicine* puts it, "Ultimately, there is no indication or rationale for the current prophylactic use of NSAIDs by athletes, and such ritual use represents misuse of these potentially dangerous agents."

Acetaminophen might be a different story, however. First of all, the drug operates differently than ibuprofen and other NSAIDs. It isn't a strong anti-inflammatory, so it doesn't have the same negative effects on training adaptation that ibuprofen does. More interesting, however, are the possible effects that acetaminophen might have if you take it before you exercise.

A study at the University of Exeter took a group of thirteen well-trained cyclists, gave them either a placebo or 1,500 mg of acetaminophen, and asked them to ride a ten-mile time trial. After taking the drug, riders were 2 percent faster than those who had gotten the placebo. But that's not all. When the riders had taken acetaminophen, they rode at a higher heart rate and produced more lactate, but had the same perception of effort as when they took the placebo. That's to say, they rode harder, but it didn't feel like it.

The lab, led by Alexis Mauger, has gone on to show that acetaminophen also provided a group of recreational cyclists with an increase in sprint performance on the order of 5 percent, mostly because repeated sprints didn't suffer as large a drop in performance as without the drug. And they have also shown that acetaminophen increases performance in hot (86 degrees Fahrenheit) conditions, by helping keep the riders' core temperatures lower due to the drug's antipyretic effects. The riders didn't just feel cooler as they exercised; their bodies actually *stayed* cooler during the effort.

These are really intriguing results, but for the time being, even Mauger advises against taking acetaminophen as a performance enhancer. As he writes in his papers, it can mask the pain that indicates an injury and potentially short-circuit the body's thermoregulatory system in a dangerous way. The potential performance benefits as well as the potential dangers— acetaminophen overdose can lead to serious liver damage and skin reactions—have even led anti-doping authorities to begin looking at acetaminophen as a drug that should perhaps be regulated.

So, when it comes to pain and recovery, feel free to take a couple of Advil or Tylenol if you sprain your ankle shooting hoops or tweak your back lifting weights. But don't rely on these drugs as a regular part of your recovery process.

Rub a Dub Dub

In the so-called "wet room" at the Australian Institute of Sport, there's a tall, narrow tub that's designed to allow an athlete to submerge himself up to the neck in water that's kept at 15 degrees Celsius. To those of us used to imperial measurements of temperature, it's hard to imagine what 15 degrees Celsius might feel like—but when I tell you it's about 60 degrees Fahrenheit, you might think it's not crazy cold.

But then you'd lower your body into the tub. Sixty-degree water is *cold*—most of us could only survive for a few hours in water at that temperature before hypothermia set in. But day after day, athletes force themselves into this tub, because some of the same things that can cause dangerous conditions like hypothermia seem to help athletes recover more quickly from hard workouts.

Cold-water immersion, as it's called, has been consistently shown to reduce muscle soreness after workouts. When a human body is placed in cold water, the heart rate and blood pressure increase to try and keep it warm. This increased circulation, along with a decrease in body temperature due to the cold water, probably helps to reduce inflammation in muscle tissues. That reduced inflammation likely helps lessen what sports scientists call DOMS—delayed-onset muscle soreness, a sensation familiar to any person who's ever overdone it at the gym and then can barely walk two days later. Cold-water immersion is especially good for athletes in hot climates or who might have elevated core temperatures, as it can help lower the body's temperature in addition to aiding recovery.

The AIS recovery team recently published a review of all the studies that have tried to measure the effect of water immersion on athletes. By looking across the studies that have shown benefits for cold-water immersion, the team came up with its current best recommendations for how to use cold baths. The most effective temperatures seem to be between 10 and 15 degrees Celsius, and the most effective treatment time is between 5 and 15 minutes—the lower the temperature of the water, the less time you'll need to spend in it. Which is good, because it's not fun to spend even 5 minutes in 10 degree C water.

But cold water isn't the only game in town when it comes to immersion therapy. There's recently been a lot of research around what's known as contrast therapy, in which you alternate between a cold bath and a tub of hot water that's at about 38 degrees Celsius. You spend a minute in the hot bath, then a minute in the cold, switching back and forth for fifteen minutes. By moving back and forth, you constrict and dilate the blood vessels, creating a sort of pumping action to help flush waste products from the muscles. The AIS recovery team has found that this sort of contrast water therapy results in less loss of muscular power after a hard workout, and faster restoration of full muscle function.

This sort of therapy is a little harder to manage. First and foremost, you need two tubs or pools that can be set up with the right water temperatures. That's why in the wet room at the AIS, the hot water tub sits right next to the cold one, allowing the athletes to easily bounce between them. It's become such a part of the life of athletes at AIS that the Australian Olympic Committee made sure it was available to their athletes in London for the 2012 Olympics. "We set up a recovery center at a school that was just outside the Olympic Village," says Shona Halson. "We had massage, hydrotherapy, active recovery, stretching. It's not exactly the same setup by any stretch, but it's got the same water temperature that they're used to, the same soft-tissue therapists they've worked with. It's all well and good to have these great training facilities, but you can't say that when you're at the games you can fend for yourself. They should be able to do all of those things at the biggest competition of their life."

Halson tells me I can get some of the same effects at home in the shower by cranking the water down to as cold as I can stand for a minute and then back to warm. Back and forth, setting up my own little contrast water therapy. Or, she says, just fill up your tub with pure cold water from the tap. "You'd be surprised how often it comes out right around fifteen degrees Celsius," says Halson. "The perfect temperature."

Newton's First Law of Sports

Like so many other areas of sports science, there's a flip side to a high-tech recovery program. There's a simple cycle that underlies all athletic training: stress, recovery, compensation. We apply a training stress to an athlete, and her body works to try and reach a state of equilibrium, growing and changing to meet the demands the training program places on it. You lift weights that are heavier than you can easily handle, and in response to that stimulus, your body works to increase muscle mass to handle the workload.

This cycle of overload and adaptation is the fundamental principle of athletic improvement, Newton's First Law of sports. But what if interventions during the recovery process after a training session have an effect on the long-term adaptation from that training? From the research, we know that recovery protocols such as hydrotherapy and compression work acutely: The next day, you feel better and are able to perform more effectively if you've worked at your recovery than if you haven't.

But long term, is it a good thing? Does a decrease in muscle soreness after a workout using compression mean that you're better off, or does it mean that you're not getting the full benefit of a hard workout? Does recovery trade long-term gain for short-term lack of pain?

The answers aren't clear. But there are analogous situations in which a measure meant to improve recovery for an athlete has been shown to actually lessen the positive effects of exercise. There's been a long tradition of athletes taking antioxidant supplements, like vitamin C, to counteract the stress that's inflicted on their cells during training. The idea is that the antioxidants limit that damage and help the athletes recover more quickly. A growing body of research, however, indicates that while antioxidants might suppress some of the markers of cellular damage, they also seem to blunt the effectiveness of physical training. Those stress-related compounds in our muscle cells are a signal to our bodies to beef up, so they can handle the load we're putting on them. Antioxidants seem to get in the way of that process, so the training doesn't have the same efficiency as it would without the antioxidants.

Whether the same holds true for recovery techniques is unclear. But there is some sense in elite circles that during periods of training, it might be better to forgo using too many recovery protocols. A scientist at the English Institute of Sport, Jonathan Leeder, has written about the possibility that decreasing stress on the body through things like ice baths will necessarily decrease the beneficial adaptations from training. Halson's lab at AIS recently looked at the issue and didn't find any negative effects from cold-water immersion for the training adaptations of cyclists. But it still might make the most sense for an athlete's progress to save recovery exercises for competitive periods, when they're focused on maintaining capabilities, not increasing them.

Some sports scientists from other countries told me they believe that's just what the UK track cycling team did in the run-up to the 2012 London Olympics—no recovery during training, and then plenty of it during competition. If that's true, it would be just one factor out of many for that team's success on the track, another of the marginal gains that the team is always chasing. But winning seven out of ten events would be a pretty good data point to include in an analysis of the optimal form and timing of recovery interventions.

10

THE NUMBERS GAME

The Power of Data to Show Us
How Our Games Really Work

In the swampy heat of Baton Rouge, Louisiana, Lolo Jones peers at a line of approaching rainclouds. The rumble of thunder carries across the track at Louisiana State University, where on this day in April 2012, Jones is training for her second run at an Olympic gold medal. Jones entered the 2008 games in Beijing as the favorite to win the 100 meter hurdles, but in the final race she clipped a hurdle and ended up finishing seventh. A lifetime of training evaporated with one error.

Jones is attended by twenty-two scientists and technicians, paid for by Red Bull, one of her sponsors. It is her seventh training session with the team, and today they've arrayed forty motion-capture cameras along the track. She's also being monitored by a system called Optojump, which measures the exact location and duration of Jones's contact with the rubberized surface on every step and after every hurdle. A high-speed Phantom Flex camera rigged next to the track will zoom alongside Jones and film her at fifteen hundred frames a second. The Red Bull team calibrates the equipment while Jones warms up.

As sports go, hurdling is incredibly technical. A runner's raw speed has to be balanced against her form and technique as she clears ten thirty-three-inch hurdles. If you're running too fast and not focusing on your form as you approach a hurdle, you can suddenly find yourself too close to the barrier as you take off, and you'll hit it, as Jones did in Beijing. Jones and her coach, Dennis Shaver, are seeking a deeper understanding of how she runs and how she might be able to adjust her technique to gain an advantage over her competition.

Just as Jones completes her warm-up, the skies crack wide open. The motion-capture session is washed out; the delicate cameras have to be hurried under a tent for protection. As Red Bull's director of high performance and former U.S. Ski Team sports science director Andy Walshe says, "We're the only people stupid enough to try and do motion-capture outside like this." But there's still the Phantom camera rig, which has no trouble handling the rain. Jones does a series of five sprints down the track, clearing the hurdles with a graceful aggression. The resulting hi-res footage is both beautiful and revealing, showing far more detail about her hurdling than the naked eye could ever see.

Jones and her coach gather with the scientists and watch the video to see how quickly she is getting her lead leg back on the ground after each hurdle. "We discovered that I wasn't kicking down my front leg as soon as I could," Jones says. "I'm just trying to get down a little sooner over every hurdle, maybe an inch closer on each one. Over the course of ten hurdles, that's ten inches, and when you're winning or losing by hundredths of seconds, that's a lot."

Richard Kirby, an engineer on the project, ticks off other discoveries the team has made through its work with Jones. They've found that she is usually fastest on her fourth or fifth trial, so Shaver increased the length of her prerace warm-up. They discovered that her left side isn't as strong and stiff as her right, which causes her to wobble slightly down the track, reducing her speed, so now she's working to strengthen that side of her body. And they found that sometimes she lands with her center of mass behind her front foot, which for a sprinter is like pumping the brakes.

The results of the research speak for themselves. After the 2008 disaster, Jones had been beset with injuries and struggled to regain her form and her confidence. In the lead-up to London, her race times were poor, but all the work finally clicked once she was there. Jones ended up running her best time in two years at the 2012 games, finishing fourth in the 100 meter hurdles and missing a medal by just a tenth of a second.

The insights that Jones's team was able to provide her aren't things that could be seen with even normal video analysis. Shaver and Jones could have trained forever without noticing them. "A lot of things you don't know, simply because you can't measure them," Kirby says. "Getting data like this puts you in a position to ask intelligent questions."

But just being able to ask those questions isn't enough. You need to be able to find the right people to look at the mountain of numbers you're generating to help you try and answer them.

Dorkapalooza

The annual MIT Sloan Sports Analytics Conference has been described by ESPN sportswriter Bill Simmons as "Dorkapalooza," a gently deflating nickname for a gathering that has grown from an MIT classroom to Boston's convention center over the course of the past eight years. It actually feels like any other trade show, a gathering of established leaders in a field as well as thousands of aspiring graduates and students hoping to find their dream job in professional sports.

But what's different about the Sloan conference is the focus, which is not on products or specific companies, although there are plenty of software vendors and inventors hawking their wares. The focus of Sloan is an idea— one that's simple and powerful, obvious and long-suffering. The idea is that analytical thinking can unlock an understanding of sports that we've missed for decades, which would provide a significant competitive advantage.

The broad history of sports analytics begins in baseball. In 1916, F. C. Lane, the editor of *Baseball Magazine*, published a story arguing that batting average was a flawed statistic that failed to account for the relative value of different hits—home runs are obviously more important than a single, but batting average views them as the same.

But batting average was easy to calculate and understand—not a small consideration when all statistics had to be recorded and calculated by hand, with calculators and computers still decades away. Lane's call for a better way to determine a batter's true value went unheeded. There were other blips of analytic thinking in the game's history, like legendary Dodgers general manager Branch Rickey hiring a full-time statistician named Allan Roth for the team in 1947, and engineering professor Earnshaw Cook publishing a book called *Percentage Baseball* in 1964. Cook's book included an analysis of the sacrifice bunt that showed it was rarely the correct strategic choice, but his analysis was largely ignored, and managers kept right on calling for bunts.

The long march of what's now called sabermetrics into baseball's mainstream really started in the mid-1970s with a night watchman at a pork-and-beans factory in Kansas. His name was Bill James, and through his *Baseball Abstract* (a book published annually), the game slowly awakened to the value of statistical analysis. James showed that there's a mathematical relationship between a team's winning percentage and its runs scored and allowed, and he created systems that quantified an individual player's contribution to a team's overall success as well as some of the first statistics that allowed defensive performance to be analyzed in the same way as offense. James was a towering intellectual figure who built a small audience over time, but he found no teams willing to work with him. (Decades later, James was hired by the Boston Red Sox, and now has three World Series rings.)

More than twenty-five years after James began his work, Michael Lewis published *Moneyball*. Lewis took readers behind the scenes with the Oakland A's, who struggled to compete with teams that had more financial resources. The A's and their general manager, Billy Beane, still managed to be successful, in large part by using strategies driven by analytics and economics. They couldn't afford high-paid stars, so the team looked for players who were undervalued by traditional understandings of the game. Home run hitters were expensive; hitters who had a knack for getting on base by drawing lots of walks were relatively cheap. Lewis's book revealed the tensions between traditional baseball thinkers and a new generation familiar with James's work who approached the game as an intellectual problem to be solved rather than a collection of traditions to be upheld.

A decade later, what was once a fierce debate in baseball between tradition and innovation is largely nonexistent. The "dorks" who populate the Sloan conference are ascendant, and the traditional old guard who once rejected them have come to understand that the combination of the quantitative and qualitative, of stats and scouting, is probably the best path to success.

But baseball is, relatively speaking, the easiest of the major team sports to analyze. At the center of the game is the one-on-one confrontation between batter and pitcher, and each play has a start and end point. Granted, there are thousands of factors to understand in those confrontations, from pitch location to defense to lineup construction. But at the end of the day, you can calculate the value of an out, or a hit, or a strikeout versus a walk.

Now, think about basketball. Instead of the static, state-to-state movement of a baseball game, you have a constant flow up and down the court. Players switch from offense to defense, from posting up to shooting outside shots. If a baseball player is a left fielder, you know the basic area he will patrol on defense. If a basketball player is a forward, he could be anywhere on the court at any time.

Understanding baseball is largely about understanding percentage and probability. Basketball is about understanding space in addition to these factors. And that could be why one of the most exciting researchers in the game is a cartographer.

"The Dwight Effect"

As a kid growing up near State College, the home of Penn State, Kirk Goldsberry was a rabid basketball fan. He wasn't quite close enough to Philadelphia to get 76ers games on TV, and so, casting about for a team, he latched on to Dominique Wilkins and the Atlanta Hawks. They might have been 750 miles away, but through the magic of TBS, Goldsberry could follow them as if he himself hailed from Georgia.

Goldsberry got a bachelor's degree in geography at Penn State, and then a master's and PhD in geography from UC Santa Barbara, where he wrote his thesis on real-time traffic maps of the Internet. All through his education, he wasn't just obsessively watching basketball; he was playing it too. "From my own experiences as a player, I know that I have strengths and weaknesses that vary depending where I am on the court, and I guessed that other players did as well," says Goldsberry. "I was always surprised that nobody had mapped players in space."

The real issue with mapping players in space was getting the data on where players were on the court. Tracking ten players who are in constant motion isn't a trivial challenge. But Goldsberry was able to track down some spatial data for the NBA—play-by-play stats that showed where each attempted score happened on the court, and the outcome of that shot. It wasn't a lot—just where the shot was, who took it, and whether it went in or not—but it was a start.

The only trick was that while the data wasn't exactly private, neither was it exactly public. What Goldsberry did was what programmers call "scraping" the data from the web (specifically, from shot charts that were displayed with the box scores for individual games). It was easy to go and look at the charts, but with a little effort, an enterprising cartographer could find the files that powered those charts, and grab all the data from them. "They were publishing these data sets, but not using them to the potential that I saw in them," says Goldsberry. "A lot of these projects come from a person getting the data and using it in a way that the original provider never thought of."

Eventually, he pulled together a database that had the spatial coordinates for every shot taken in the NBA from 2006 to 2011—more than seven hundred thousand of them. Once he had that data, Goldsberry the cartographer and Goldsberry the hoops junkie were able to get to work. "Basically, I applied all these things I had studied so hard to this new data set," he says. "I wanted to find a way to get this data to sing a new song, to tell us things like where Kobe is good and where Kobe is bad. Instead of stats—using letter and numbers—showing it visually was a way to communicate to players and fans and the media."

What Goldsberry came up with was a system he dubbed CourtVision. He divided the areas of the court where players actually shoot—basically, from just outside the 3-point line and closer—into 1,284 one-square-foot areas. Here are the results from that first data set, showing how frequently shots were taken from each area, and how often they were made.

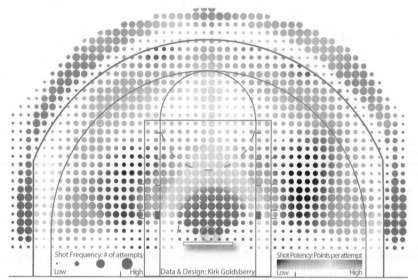

Beyond the aggregate, Goldsberry could generate maps for individual players that showed, from each of those areas, how many shots he took and how effective those shots were in terms of points scored. Suddenly, it was easy to see the differences in players. Ray Allen had several deadly hot zones from 3-point range, and he barely attempted any midrange jumpers. Kobe Bryant took lots of shots from all over the court, but there were places you'd much rather he shot from (like the left baseline, because he struggled to convert from there). It was an instant visual signature of a player's offensive game, easy to read and understand, yet the more you studied it, the more insights it revealed. Goldsberry presented his work at the 2012 Sloan conference, and the basketball world basically freaked out. For the first time, fans could get a view of the types of shots that their favorite players took, and the relative value of those shots. It gave team owners and management a powerful tool to evaluate players, to make sure that they were efficient and that their style fit in with a team's philosophy.

"I had people like [Dallas Mavericks owner] Mark Cuban and [San Antonio Spurs GM] R. C. Buford coming up to me and expressing interest in this project that I had just been doing in my spare time as a professor," says Goldsberry. "It was sort of a moment of 'Oh my god, if I do this right, I might be able to go turn this into something that's bigger than just a thing I do on nights and weekends.'" He went home that night and started a blog.

One of the people who were intrigued by Goldsberry's work was Brian Kopp, an executive at STATS in Chicago. STATS was started in the 1980s by a group of baseball researchers who were trying to gather the best statistical information they could about the game. The company has since grown into a behemoth that provides statistical information and content about nearly every major professional sport in the United States, and many more around the world. Kopp asked Goldsberry if he would be interested in working with a new kind of data that STATS was developing for NBA games, using a system called SportVU.

SportVU builds on technology that was actually developed by Israeli scientists for tracking missiles in defense applications. In 2005, the Israelis adapted their optical tracking technology to sports, first focusing on soccer. In 2008, STATS bought SportVU and decided to use the system for basketball.

The physical technology behind the system isn't all that mind-blowing. A set of six cameras is installed in the arena above the court. These cameras are all connected to a central computer, which takes the computer vision information from each one and combines the data. That allows every object on the court, from the players to the ball to the officials, to be plotted in three dimensions, twenty-five times a second.

That positional data can then be linked with play-by-play information to reveal where and how a player is moving, and how that affects the outcomes during a game. Want to know how far a player runs during a game? No sweat. Wondering who the most efficient passer is on your team? Easy. Trying to identify which shots are most effective over the course of a season? You got it. But it can get massively more granular. How does your pick-and-roll efficiency compare with the league average when you start the move with less than fifteen seconds left on the shot clock? SportVU data can reveal the answers.

Through the end of the 2012–13 season, any individual NBA team who wanted this information had to pay roughly a hundred thousand dollars for the installation of the camera system in their arenas. Only fifteen teams had done so, which meant that there were huge gaps in the data; roughly half of the games played weren't captured. But the value of SportVU data has become so clear that the NBA signed an agreement in September 2013 to install the system in every arena in the league.

For Goldsberry, SportVU was the Holy Grail. "Brian called me and was basically like, 'Do you want to play with this data?'" says Goldsberry. "I had the good fortune to get access to that data when very few people outside of the NBA had seen it, when there hadn't been people with the spatial analysis and analytic skills looking at it."

One of the most exciting possibilities that spatial data presents is truly understanding the role of defense in the NBA. We've traditionally relied on simple counting stats such as steals and blocks to capture the defensive value that a player brings to the court. Using SportVU, we can have a much more sophisticated view of defense, giving us a clearer picture of which players actually help and hurt their teams the most at the defensive end. You can look at the data and understand the best way to play defense against a pick-and-roll, or which players are especially good at getting into passing lanes to disrupt the offense. You can even see which players change the opponent's offense just by being on the court.

That was the promise of the presentation that Goldsberry gave at the 2013 Sloan conference. His talk was titled "The Dwight Effect," after then–Los Angeles Lakers center Dwight Howard, and the room was packed, not just with fellow researchers but also with executives from around the NBA.

Goldsberry started by observing that the area right around the basket is the most important real estate on the court to defend, as it's the area where offenses make the highest percentages of their shots. After identifying this, Goldsberry looked at what happened on shots that were taken against a defender who was within five feet of the basket. How could defenders stop opponents from scoring there? "We assert that 'dominant' interior defense can manifest in two ways," wrote Goldsberry in a published version of the research. "Reducing the shooting efficiency of opponents, and also reducing the shooting frequency of opponents."

It turns out that there was a huge difference between players' abilities to reduce shooting efficiency. Indiana Pacers center Roy Hibbert and Milwaukee Bucks center Larry Sanders were stars by this metric. The average NBA defender allowed a shooting percentage of 49.7 percent in these interior situations; Sanders and Hibbert held opponents to 38 percent shooting, both by blocking a lot of shots and by forcing opponents to alter their shots. That's a massive boost for their teams' defense. On the flip side, Luis Scola, then of the Phoenix Suns, and David Lee, of the Golden State Warriors, are defensive disasters, allowing shooters to make 63 and 53 percent of their shots respectively. For a player like Lee, who was an all-star in 2013 (thanks to his gaudy offensive production and rebounding numbers), this sort of analysis might be bad news when it comes to his next contract negotiation. It was the opposite for Sanders, who signed a $44 million contract in August 2013, in part because of his now-demonstrated defensive value.

Then there's the other half of Goldsberry's equation: reducing the shooting frequency of opponents near the basket because of the fear of having the shot blocked or altered. By comparing the average rate of shots to the rate when specific defenders were guarding the area, Goldsberry could calculate which defenders prevented the shots from happening simply through their presence. The leader was Dwight Howard, who caused teams to shoot 9 percent fewer shots just by being there as a defender. As Goldsberry writes:

This is what we call the "Dwight Effect"—the most effective way to defend close range shots is to prevent them from even happening. Although Howard does not lead the league in blocks, he does lead the league in "invisible blocks," which may prove to be markedly more significant. When Howard is protecting the basket, opponents shoot many fewer close range shots than average, and settle for many more mid-range shots, which are the least productive shots in the NBA.

One of the NBA executives in the crowd at Goldsberry's talk at the Sloan conference was Daryl Morey. Morey is the general manager of the Houston Rockets, where he's turned the organization into one of the most forward-thinking in the league, investing a great deal of time and energy in analytics and sports science. An alum of the MIT Sloan School of Management, which puts on the annual conference, Morey actually cofounded the event and continues to serve as one of the cochairs.

Maybe it's a coincidence. Maybe it's not. But it's worth noting that four months after watching Kirk Goldsberry demonstrate some of Dwight Howard's hidden defensive value, Morey signed Howard to a massive contract to lead his Houston Rockets in their title chase.

Taking the Lab to the Field

Camera-based systems like SportVU are great technical accomplishments, and have the advantage of not requiring any direct interaction with athletes. But there's another way to measure what's happening with athletes on a field, and that's by using a piece of electronics attached to an athlete that captures information about his or her movements. Tracking our movements and our data has been a booming industry in the past few years, as devices like the Nike+ FuelBand, the Fitbit, and the Jawbone Up offer regular people a way to measure their activity and motion. That boom in sensors, not to mention the phones that many of us carry (which are packed with accelerometers and gyroscopes), has led to a whole new generation of people trying to optimize their lives through data—what's been dubbed the Quantified Self movement.

Peter Drucker's now-famous quote, "What gets measured gets managed," is especially true for sports. One of the things that draw many people to sports is that the final evaluation of your performance is so clear-cut—just look at the scoreboard or the results sheet. Every day most of us go to work and do our jobs the best way we know how. But there's no immediate sense of how well we've done. When I write a good sentence (as I hope I do from time to time), I have a sense of satisfaction, but there's no score, no objective measurement.

Sport has the opposite problem: The focus on the competitive outcome can sometimes overshadow real progress and improvement. But the key to understanding that progress is to collect data on it. Athletes have long been at the forefront of this movement, from keeping training logs to breaking down film, but now, with the wealth of technology we have available to collect and analyze information, the importance of just gathering and managing data has never been higher.

Without data like the numbers you can get from a Fitbit or a Jawbone Up, you'll never know exactly what's going on. If you don't track your workouts, if you're not testing your fitness once in a while, you won't know if you're wasting your time in the gym or if you've found a routine that really does help you. You'll never know the results of the ongoing experiment that all of us are engaged in when it comes to our bodies and fitness, and you'll never have any basis to make informed decisions.

The elite version of a Fitbit comes from an Australian company called Catapult Sports. The company is the outgrowth of what is called a Cooperative Research Centre, an Australian governmental program that is meant to help spur technology growth in the country. The founders of Catapult, two mechanical engineers named Shaun Holthouse and Igor van de Griendt, were tasked with finding new ways of using microtechnologies, and they soon launched a project with the Australian Institute of Sport.

The resulting gizmo was a small GPS unit called Minimax. It allowed the AIS researchers to capture data on athletes and what they were doing outside a lab setting, so it was more true to the actual demands of a sport. After working with AIS to refine the device, a separate company, Catapult, was spun out to sell the system in 2006.

Today, Catapult's tracking unit, known as the OptimEye S5, contains fifteen different sensors, according to Adir Shiffman, the company's chairman. They range from GPS, gyroscopes, and accelerometers to magnetometers and heart rate monitors. The device is a portable sports lab; it fits into a sort of harness that holds it between an athlete's shoulder blades. The athlete goes about her normal routine, and the OptimEye constantly beams information back to the coach and analytics staff in real time, tracking how far and how fast she's running, the stops and starts, how quickly she's turning and cutting, accelerating and decelerating.

Catapult had its first big success in its home country. "Australia's the clear world leader in sports science," says Shiffman. "And the biggest sport here is Australian rules football." Aussie rules is mostly familiar to U.S. audiences as one of ESPN's early programming staples, but it's actually a very interesting game. There are tons of players, a huge field, lots of running and hitting, and very interesting tactics. Catapult began to work with several teams, and now seventeen of the league's eighteen teams are clients.

One way those teams use the system is to make decisions about substitutions during games. Aussie rules allows players to swap in and out almost constantly, so one key decision for coaches is which players to bring off and on, and when. Catapult's software attempts to help with those decisions. Using an iPad on the sideline, a coach can see what's known as the player load for each man on the field—how far they've run, how much rest they've had, how much contact, and so on. Crunching those numbers, the software can also see when a player's performance starts to slip due to too high a load, and suggest that he come off for rest. "A lot of the decision making for interchanges is happening based on that data," says Shiffman. You don't need to ask a player if he's tired—you know. Imagine how game changing this simple piece of knowledge could be for any sporting event, whether it's the Super Bowl or a weekend flag football game with a bar tab on the line.

After Australia, the system started to spread around the world, first to the UK and Europe—mostly to soccer clubs—and now to the United States ("This is one of the rare occasions where the U.S. represents an emerging market," says Shiffman.). The company is working with an increasingly large number of college programs, as well as at least seven NFL teams and eight NBA franchises. In fact, Mark Cuban, the owner of the Dallas Mavericks, invested in Catapult after the team started using the system. None of the U.S. pro leagues allow the product to be used in regular season games yet, so the focus is really on two things: quantifying training load to ensure that athletes don't become fatigued, and tracking rehab to know that an athlete is ready to handle the actual conditions he'll face in a game.

If you're tracking a player's performance in every practice, you have a baseline set of data on what he does. Imagine you have that information, and one day he's unable to accelerate in the way he normally does. Instead of having to guess what's going on, you can look at the data and talk to the player. Maybe he has a physical problem that's not keeping him off the field yet, but could if it's not treated. If you can flag a couple of potential injuries like that, the system's annual six-figure cost is well worth it, given the amount that pro athletes are paid.

Shiffman recounts how the company got one of its newest clients, the New York Giants. "Usually, our first conversation with a team comes from someone like a strength and conditioning coach," he says. "With the Giants, the conversation started at a much higher level. They realize that this system can deliver a material competitive advantage to the team." It seems like just a matter of time before leagues start to allow these devices in games, and teams like the Giants are looking to get a head start on collecting and understanding the data.

Candid Cameras

Sportvision occupies a squat office building in Mountain View, California, not far from Sparta Performance Science, and just north of NASA's Moffett

Field. Pulling into the parking lot, you'll see little more than a sign marking your arrival at the nerve center of the modern presentation of sports on TV.

The yellow line that shows football fans how far a team needs to go for a first down? Sportvision invented that. The pitch tracking in baseball that shows whether a pitch was a ball or a strike? Sportvision. The pointers and graphics in NASCAR broadcasts that show the exact gap between cars, and their speed in real time? Even the much-reviled glowing puck that Fox used for professional hockey games—all Sportvision. The walls of the office are lined with framed versions of the patents the company holds for all of the tech behind these TV tricks. A number of them are relatively straightforward in concept, if not execution. To put the yellow line on the field, the company uses a special attachment for the cameras shooting the game that allows graphics to be inserted into the shot. That sounds easy, but it took years to develop.

The company's system of tracking baseball pitches is considerably more complicated. Pitchf/x, as it's known, relies on a set of three cameras positioned around the ballpark (it's installed in all thirty major league stadiums, as well as forty minor league parks). The cameras are all calibrated with Sportvision's software to track ball-shaped objects in their field of vision. Five times a second, the cameras capture the position of the ball, and from that data, the complete flight can be calculated—the speed, the break of the pitch, where it crosses the plate. Seconds after the pitch is thrown, the results can be shown during the game broadcast, and the pitch is shown on MLB's At Bat app.

Since the debut of the system in 2006, Pitchf/x has recorded data on millions of pitches. Baseball has always been fascinated by stats, but this was a new level of granularity—not just the outcome of an at bat, but exact data about every pitch thrown. It's become a treasure trove of information for baseball analysts, both pros and amateurs; what's funny is that the data wasn't even supposed to be available to data scientists outside of the teams. It's not publicly available, at least officially. But Major League Baseball posts the data on the web so various apps can access it, and fans soon discovered that they could scrape the data (like Goldsberry did with the NBA data) and use it themselves.

The result was a boom in analysis. As Hank Adams, the CEO of Sport-vision, puts it, "This wasn't an amount of data that you could just look at. You needed to be a real data scientist, and there are a lot of smart fans who are data scientists." Soon, bloggers and scientists were crunching through the numbers to find patterns that had long been hidden from view. Questions like what pitch a certain pitcher tends to throw in certain situations or that pitcher's velocity at the start of the game as opposed to the end became answerable. These bloggers were so successful that many of them were hired by the teams themselves to do analysis.

Sportvision develops its systems with a focus on broadcast and media, but the data is even more valuable to teams and coaches. The company now offers other products based on the same technology, including a system called Fieldf/x, which shows what all the defensive players on a baseball team do when a ball is hit, tracking each player's movements and actions. This newer system requires a separate subscription, which not all major league teams have been willing to pay for. And sometimes, that has led to a competitive disadvantage. (At the 2014 Sloan Sports conference, Major League Baseball Advanced Media announced its own player and ball tracking system that uses cameras as well as radar to capture the action on the field. Major League Baseball plans to evaluate the two systems head to head.)

Adams recalls one day when he got a call from the Los Angeles Angels front office. The team had signed superstar first baseman Albert Pujols before the 2012 season, and over the course of the year, they had noticed that opponents were playing defense in a very different way when Pujols was at the plate. Some used a traditional alignment in the field, while others shifted the defense strongly to one side. The teams that shifted against Pujols all had one thing in common: They subscribed to Fieldf/x. The Angels realized that the data must show something conclusive about Pujols. Their only solution—a good one for Sportvision—was to subscribe to the data package to find out exactly what that data was.

Another intriguing possible use of Pitchf/x data is in injury prevention. After all, teams spend millions of dollars on pitchers each season, and keeping your best players healthy is paramount. By tracking the performance and, most important, the changes in performance of its pitchers, a team might be able to see signs of fatigue and possible injury—or even a change that leads to a competitive disadvantage—before it becomes an acute situation.

As an example, Adams pointed to the case of San Francisco Giants pitcher Tim Lincecum. Lincecum won the Cy Young Award both of his first two full seasons in the majors, and made the all-star game four consecutive seasons. But in 2012, he began to struggle for the first time in his career. The Sportvision team wasn't shocked. "Our system noticed a difference in his mechanics almost immediately," says Adams. "His release point was different than it had ever been. It was so different, we wondered if our cameras got bumped out of alignment." Differences that would be hard for even the most seasoned baseball coach to pick up stick out like a sore thumb in the data.

The New Competitive Advantage

Every conversation about the use of statistical analysis in sports returns, as if drawn by its inescapable gravity, to *Moneyball*. Part of that is because it's just such a terrific book, and Billy Beane is such a great character. And Michael Lewis's storytelling prowess made it easier to understand the stats.

But I think we've missed a key lesson in the success the Oakland A's had by implementing an analytical focus. Much of the statistical information that Beane used to transform the A's had existed in one form or another for years. He did synthesize some of it, yes, but more important, he was able to operate his organization along those principles. That's to say, the competitive advantage didn't come from a novel theory of the game; it came from being able to move his franchise along an analytically driven path more quickly than his competition.

As new technologies start to generate terabytes of data about players and tactics, that next great competitive advantage won't be about acquiring the data; it will be about understanding it. It's a good time for all those data guys at the Sloan conference, because they're the ones who will be called upon to tease meaning from the numbers. "When you get down to the troops on the ground, it's not an exaggeration to say that eighty-five percent of the teams don't know what to do with this data," says Kirk Goldsberry about the NBA SportVU information. "The idea that this is going to revolutionize the NBA, well, I'm not sure that's true unless teams awaken really quickly to things like machine learning and data visualization. There are all of these things that they don't know how to do."

It's that 15 percent of teams who *do* know what to do with the data that could have the sort of competitive edge that Billy Beane found. "For the smart teams, it will become a bigger and bigger advantage," says Goldsberry. "They already have video analysis and scouting. The data gives you a sort of third eyeball on things, to characterize performance, the value of players, the effectiveness of your strategies and your opponents' strategies."

Consider Kevin Youkilis, the minor league baseball player who Billy Beane memorably tried to get in a trade. Beane, who saw Youkilis snatched away in the draft by the Boston Red Sox, called the player the Greek God of Walks, for his uncanny ability to draw walks and get on base. (In those days, on-base percentage was undervalued by most front offices.) Beane was right to covet Youkilis, who has gone on to a very solid professional career, making three all-star teams and helping the Sox win two World Series. Goldsberry wonders who will be the Kevin Youkilis of the NBA. "Who becomes that player who's emblematic of the analytic movement?" he asks. "Maybe we're seeing it already. Daryl Morey likes what he calls 'attack guards,' players like James Harden and Jeremy Lin who are famous for attacking the middle of the floor and getting fouled, guys who don't shoot many two-point jump shots. But I think you'll see teams start to understand the game differently and make what might seem like head-scratching signings and trades."

At some point, one of those players—someone whom the conventional wisdom didn't value but analytics did—will make it to the upper echelon of the game. That's when we'll have that player, the story that will drive home to every team the importance of the data-driven future they're already living in.

11

ATHLETE'S LITTLE HELPER

From Champions to Cheats and Back Again

The 1988 Olympics in Seoul got off to a good start for Dick Pound. As vice president of the International Olympic Committee, he had helped return the games to glory after the ruinous Cold War boycotts in 1980 and 1984. Millions of fans were in Seoul, and—more important for Pound—billions more were watching in 160 countries across the globe. Pound was in charge of TV rights and had brought in a record $403 million from broadcasters to air the 1988 summer games. To top it off, Pound, a Canadian, was in the stands at Olympic Stadium when his countryman Ben Johnson sprinted his way to a gold medal in the 100 meter final. Johnson's time of 9.79 seconds shattered his own world record, and in beating America's Carl Lewis, Johnson confirmed his position as the fastest man on the planet.

The day after Johnson's victory, Pound was still glowing, soaking up congratulations at a lunch with Olympic sponsors, when Juan Antonio Samaranch, president of the IOC and Pound's mentor, burst into the room. Samaranch, known for his aristocratic manner, was in an uncharacteristic panic.

"Dick," Samaranch said, "have you heard the news?"

"What is it?" Pound asked. "Somebody died?"

"No, no, no, it's worse," Pound recalls Samaranch saying. "Ben Johnson has tested positive." More precisely, Johnson's postrace blood sample showed evidence of stanozolol, an anabolic steroid used to boost lean muscle growth.

The scandal threatened to unravel the work Olympic officials had done to recover from the stain of the two boycotted games. A quick, resolute decision from the IOC medical commission was in order. Desperate to save their medalist's reputation, Canadian officials asked Pound, who was an experienced attorney, to represent Johnson at the hearing, which would determine whether he would keep his medal or be booted from the Olympics.

At that point, Pound was the heir presumptive to Samaranch as head of the world's most powerful sporting organization. Before he put his name and reputation on the line, he wanted to talk to Johnson. Pound pulled him aside in the only private space they could find: the bathroom in Pound's hotel suite. "Ben, are you on anything?" he asked. Johnson looked Pound straight in the eye. No, he said. He had no idea how the drugs could have ended up in his system.

Pound took the case. In the hearing, he argued that someone had sabotaged Johnson's sample or that it had been accidentally contaminated. But the scientific evidence was overwhelming. Blood tests showed that not only did Johnson have stanozolol in his system, but that his adrenal function was suppressed, indicating long-term steroid use. This was no glitch.

The verdict was swift: Johnson was stripped of his medal and suspended for two years. Three days earlier, he had been a champion. Now he was a cheat. Other Olympians had tested positive, but never a gold medalist in the premier event of the games. It was undeniable that drugs had permeated sports at the highest level and that sporting officials were lagging far behind.

The case also marked a turning point for Pound, who overnight went from romantic to cynic. "Most athletes, when they're caught, lie," Pound told me when I was writing about the situation years later for *WIRED*, the disappointment still fresh on his face. "Their coaches lie. The people around them lie. They just deny, deny, deny."

A Dent in the Armor

When I started writing this book, I wasn't planning to include a chapter on performance-enhancing drugs. I had written about doping for *WIRED* when I profiled Pound. Looking at that story today, I see a writer who is groping for a way to continue to believe in sports, to set aside the suspicion and cynicism that I attributed to Pound. I wanted to look at doping, but I didn't want to *confront* it.

That story recounted how Pound had left the IOC to be the first head of a new organization called the World Anti-Doping Agency, or WADA. The idea was to create an independent international authority to regulate and police drug use, keep it separate from the IOC and the individual sports' governing bodies, and make governments part of the process so the organization could use their powers of arrest and subpoena.

It was an elegant antidote to two great stumbling blocks in the fight against drug use. The first issue was that each sport had its own list of outlawed substances, creating confusion about what was legal and what was banned. WADA would cut through the clutter by publishing a single, unified list of banned substances to be adopted by all Olympic sports. From steroids to stimulants, from hormones to narcotics, every athlete in the world would be held to the same standard.

More crucially, WADA would establish a clear and precise process for all drug testing to follow. Each WADA-accredited lab, whether in Bangkok or Bogotá, would follow the same procedures in handling and processing urine and blood samples from athletes. If a foreign substance was found in testing, the code laid out in exquisite detail what would happen next: A so-called B sample (taken at the same time as the A sample) would be tested to confirm the result. If B was negative, the investigation would end and the athlete would be exonerated. If both samples were positive, the lab would forward the results to the anti-doping agency in the athlete's country.

So, for example, the U.S. Anti-Doping Agency would be apprised of the positive test of an American athlete, and it would be responsible for bringing formal charges against him, leading to a hearing and subsequently a decision. Everyone involved, from the athlete to the national anti-doping agency to WADA itself, would have the right to appeal the decision to the Court of Arbitration for Sport, which serves as an international arbiter for sporting issues and has the final word. The process would be clear, fair, and beyond reproach.

The World Anti-Doping Agency came into existence in 1999; by 2003, the code had been adopted by all Olympic sports. Objectively, WADA has been a great success. Just the act of organizing all anti-doping efforts under the same set of rules has been hugely beneficial, eliminating confusion over which drugs are banned in which sports.

When I was writing about Pound in *WIRED*, he was in the midst of a battle with Floyd Landis, the American cyclist who had won the Tour de France in 2006 but was then disqualified after he was found to have used testosterone. Landis's ride in that tour had been operatic—he'd had a disastrous day that appeared to cost him the race, and then he'd set himself up for the win with an audacious, seemingly futile attack.

In my reporting for the story, I interviewed Landis at my office. He limped in, still recovering from hip surgery. I liked him a lot. He seemed refreshingly free from the artifice you often encounter in professional athletes. His critique of the anti-doping lab that processed his tests and its shoddy work was compelling. I left that interview thinking that it didn't matter, in some way, if Landis had doped or not—the system and the process were so messed up that there was no way they should be able to destroy someone's career and life. I came to believe that Pound was too vocal, too quick to judge, and too loose with pronouncements of guilt. I still think some of that is true, but of course, now we know Pound was completely right.

I had actually tried to get another writer to do the story about Dick

Pound and WADA, someone who had spent a lot of time with Lance Armstrong. When I gave this writer a call to see if he was interested in doing the story, he sighed and told me, "I can't plunge myself into that world again."

But then that world was cracked wide open, especially in the realm of cycling. Books like Tyler Hamilton and Daniel Coyle's *The Secret Race* broke the silence that had surrounded doping in cycling. Landis finally admitted to using drugs, and then worked with the U.S. Anti-Doping Agency to bring down the largest target of all: Armstrong. By the time of Armstrong's confessional interview with Oprah Winfrey in January 2013, the whole of the edifice—the threats and intimidation, the endless lies— was laid bare. It was sad and disappointing, and I was tired of hearing about it and talking about it. Like that writer whom I had contacted, I didn't want to plunge myself into that world again.

The sad truth is that people cheat, and not just in sports. Don McCabe, a Rutgers professor, has been studying academic cheating in high school and college for decades, and his findings paint a grim picture. According to his research, 95 percent of high school students admit to some form of cheating, while more than 80 percent of college students say they have engaged in "serious cheating" such as cheating on a test or plagiarism. People cheat for all sorts of reasons, from thinking that the rules they're breaking aren't important to thinking that cheating is so prevalent that they have to join in to keep up. Perhaps it's crashingly naive to think we might have clean sports—why should we expect that arena to be more moral than the rest of our lives?—but if we are ever going to eliminate doping, we have to understand the thought process behind it.

The Goldman Dilemma

It's not surprising that athletes are willing to push the boundaries of morality and science in their quest for sporting success. After all, they have likely spent much of their lives in single-minded pursuit of a nearly impossible goal, and the promise of an effective way to help get there must be seductive. Add in the culture of paranoia and suspicion that can build up around performance-enhancing drugs ("everyone else must be using") and it's easy to see how an athlete can make the decision to start down the wrong path.

But just how many athletes would make this choice? And would they make it if the consequence of taking the drugs was death? That was the question Dr. Bob Goldman set out to answer with a survey that he first published in his book *Death in the Locker Room*.

Goldman was a weightlifter as well as a doctor, and he had seen two of his friends die after taking anabolic steroids to increase their performance. He decided to ask elite athletes a simple but fraught question: If there was a drug that would allow you to win every event you enter, but would kill you in five years, would you take it?

More than half of the athletes said they would take the drug. Five years of sporting invincibility would be worth paying for with their lives. Goldman did a similar survey every two years for more than a decade, and the answers were always similar: Half of the athletes, many of whom were training for the Olympics or other professional leagues, would give up their lives to stand at the top of the podium.

The question became known as the Goldman dilemma, although it didn't seem like much of a dilemma at all. A variation of the question, in which athletes were asked whether they would take a banned performance-enhancing drug that would ensure a win if they knew they would not be caught, showed even starker results: Of the 198 athletes Goldman asked in 1995, 195 said that they would take the drug.

That's a massive percentage, and much higher than the numbers of athletes who do get caught. In 2012, WADA-sanctioned labs performed 267,645 tests on the primary samples of athletes, and 4,723—1.76 percent—of those showed, in WADA's legalistic syntax, an "adverse analytical finding" or an "atypical finding." An adverse finding means the sample shows the presence of a banned substance or evidence of a prohibited training method, while an atypical finding suggests further investigation is necessary.

If you approach the problem of doping like a behavioral economist, two ways to put a stop to cheating come to mind. You can lower the incentive to cheat, making the reward of breaking the rules not worth the effort. Or you can increase the ramifications of cheating, with possible penalties or outcomes so severe that the reward doesn't seem good enough. What's most striking about the Goldman dilemma isn't that athletes would cheat. What's striking is that the athletes said that they'd be willing to cheat when the consequence of that action—death—would be so great.

Nonathletes don't have the same view of the stakes. In 2009, Australian researchers led by James Connor posed the Goldman dilemma to a group of regular citizens, and the difference in response was stunning. Rather than about half of the group saying they'd trade their lives for a gold medal, only two of the 250 respondents, just 0.8 percent, said they'd do so. Now, maybe part of that is just because the general population doesn't think sports are that important, but it's hard to explain away the huge difference as just being about priorities.

It could be that there's been a real change in people's opinions about drug use in sports. For the past decade, WADA has been trying to convince athletes that the importance of fair play trumps the benefits of using drugs. And high-profile doping cases such as the suspensions and bans of athletes from Marion Jones to Lance Armstrong have shown that drug testing is more effective than it was back when Goldman started asking athletes about their attitudes.

There's some evidence that this different environment has led athletes to change their attitudes about doping. When Connor tried to replicate Goldman's survey at an elite Canadian track meet in 2012, he discovered that nowhere near half of the athletes said they'd take the Faustian bargain. In fact, only two of the 212 athletes interviewed said they would—about the same low rate found in the general population study.

It's far too early to proclaim success in changing the attitudes of athletes toward drugs; too many athletes still get caught for us to think that somehow we've bled the desire for pharmaceutical success from sports. But Connor's survey does give some hope that a new generation of athletes are able to recognize that their lives are more important than winning.

Just because many athletes aren't as willing to accept the Goldman dilemma doesn't mean that they're necessarily competing clean. A UK study used an online survey to understand the prevalence of doping and the attitude of athletes toward it. In the survey of 729 competitive athletes ranging from the club level to international elite athletes, 2.3 percent admitted that they were currently using performance-enhancing substances, and 4.5 percent said they had previously used them.

One other interesting finding of this study was that the athletes who seemed to have the most favorable attitudes toward doping were those who were competing at the university and national level. Elite athletes are under the watchful eye of WADA and their national drug testing agency, even if the quality of those agencies varies. National-level athletes are close enough to the elite level that they might feel that drugs could get them over the hump to that next level of competition, and since they're less likely to be tested, perhaps they perceive the risks to be lower.

There's an ironic coda to Goldman's story. The man whose early work helped illuminate some of the dangers of steroids and how athletes viewed and used them has gone on to cofound the American Academy of Anti-Aging Medicine, which promotes the use of testosterone and human growth hormone to slow the body's decay over time.

Not Just Vitamins

It had an innocuous, bureaucratic-sounding name: State Plan Research Theme 14.25. But in reality, it was the name of the largest sporting fraud ever committed, dwarfing later doping scandals such as the BALCO case, which took down Barry Bonds, Marion Jones, Tim Montgomery, and dozens of others. In 1974, the Central Committee of the East German Socialist Party passed a bill decreeing that doping should be part of the preparation of all athletes for international competitions. This systematic doping of thousands of athletes in 1970s and 1980s East Germany was monitored by the Stasi, the country's secret police, through a network of coaches, doctors, and informants.

What's amazing is that during this entire history of doping, East German athletes only tested positive for steroids twice in international competitions. How did they get away with it? They enlisted the help of the country's anti-doping laboratory. Before an athlete traveled to an event like the Olympics, the administration of the drugs would stop so they wouldn't test positive. To make sure their blood was clean, East German officials ran their own drug tests. If an athlete's body hadn't cleared the remnants of their doping program, the athlete was simply held out of the international competition.

State Plan Research Theme 14.25 was a horror; many athletes were given performance-enhancing drugs without their knowledge, instead being told they were vitamins. When the program's existence was revealed after the fall of the Berlin Wall, it resulted in story after story of athletes who had suffered physically and mentally for the state.

Part of the challenge we face in talking about drug use in sports is knowing just how effective the drugs are—you can't really do good scientific studies on their effects on competitive athletes, for obvious reasons. For the most part, the real numbers on doping's effectiveness lie in the shadowy world of the athletes, coaches, and doctors who have prescribed and tracked the use of these drugs. Perversely, the existence of State Plan Research Theme 14.25 has given us one of the only sources of data we have on exactly how effective doping is for elite athletes. Werner Franke, a biologist, and his wife, Brigitte Berendonk, a former Olympic discus thrower for East Germany, were able to save some of the meticulous records kept during the program, and later went on to write a remarkable scientific paper, "Hormonal Doping and Androgenization of Athletes: A Secret Program of the German Democratic Republic Government." It's the rare work of research science that reads like a Cold War thriller, a dystopian look at how far a country would go for sporting success.

The program centered on the use of a drug called Oral-Turinabol, a steroid that had androgenic effects—that is to say, it didn't just increase muscle mass; it also increased male sexual characteristics. So it's not a surprise that the East German program was particularly effective for female athletes. By 1977, Manfred Höppner, the deputy director of the German Sports Medical Service, reported to the Stasi:

> At present anabolic steroids are applied in all Olympic sport-ing events, with the exception of sailing and gymnastics (fe-male) . . . and by all national teams. . . . The positive value of anabolic steroids for the development of a top performance is undoubted. Here are a few examples. . . . Performances could be improved with the support of these drugs within four years as follows: Shot-put (men) 2.5–4 m; Shot-put (women) 4.5–5 m; Discus throw (men) 10–12 m; Discus throw (women) 11–20 m; Hammer throw 6–10 m; Javelin throw (women) 8–15 m; 400 m (women) 4–5 sec; 800 m (women) 5–10 sec; 1500 m (women) 7–10 sec. . . . Remarkable rates of increase in performances were also noted in the swimming events of women. . . . From our experiences made so far it can be con-cluded that women have the greatest advantage from treat-ments with anabolic hormones with respect to their performance in sports.

This is just basic hormonal science—women don't have much testos-terone or other androgens, so a drug with these androgenic effects would produce some very large improvements. But these gains didn't come with-out a price. In the same report, Höppner wrote:

> In numerous women the prevailing administration of anabolic hormones has resulted in irreversible damages, in particular in the swimming events, for example signs of virilization such as an increased growth of bodily hair (hirsutism), voice changes and disturbances in libido. The effect on the sexual drive was relatively strong in some women. This resulted in special prob-lems, particularly in training camps where the "official" male partners of these women were not present.

How pervasive were steroids at the upper level of female sports in this time period, not just in East Germany, but in other countries? It's hard to give a clear-cut answer, but we can draw some inferences when we look at the list of world records in women's track and field.

Out of the nineteen world records in the track events contested in the Olympic games, from 100 meter dash to the 10,000 meters, as well as the jumping and throwing events, ten of the records that still stand today were set in the 1980s. Six of those were set in 1988 alone. (I've removed the hammer throw and javelin from this analysis; hammer wasn't an event for women until 1994, and the javelin was redesigned in 1999, making it difficult to compare those marks.)

Seven of those records set in the '80s are held by athletes from the former Soviet bloc—countries that we know had systematic doping programs. Two more records were set by Chinese runners in 1993, the same year thirty-one Chinese athletes tested positive, although the two athletes who still hold world records weren't among those busted.

And then there are the two records held by American Florence Griffith Joyner in the 100 and 200 meters. Prior to her breathtaking 1988 season, Flo-Jo had been a very good sprinter—but in 1988, she was otherworldly. Her previous best in the 100 meters had been 10.96 seconds, but then she dropped nearly half a second in a year. In the 200, she had run 21.96, but improved by 0.62 seconds in 1988. Griffith Joyner streaked across the track and field world in 1988, and then retired—just before random, out-of-competition drug testing started in 1989.

Griffith Joyner never failed a drug test. But her times fail the smell test, especially once they're put in the context of other athletes' performances from the mid- to late '80s that were almost certainly chemically enhanced. Griffith Joyner maintained that she didn't use performance-enhancing drugs throughout her retirement; in 1998, she died in her sleep from a seizure at age thirty-eight.

The upshot is that it's very, very difficult to set a world record as a female track athlete—the average age of all these records is nineteen years. Many competitors today are chasing records that are older than they are, and given that many of those records aren't even under threat (judging from recent performances), it could be at least another generation before we see them broken.

When we look at the effects of other drugs, there's not a lot to go on. A study was published in 2013 by Yannis Pitsiladis as part of a research project funded by WADA to understand the effectiveness of drugs in sports. In the study, Pitsiladis's group gave research subjects erythropoietin (EPO), a favorite drug of endurance athletes like Lance Armstrong in the late 1990s. EPO is a hormone produced in the kidneys that controls the body's production of red blood cells. If the kidneys detect that there's a reduced level of oxygen delivery to them, they produce more EPO, which then signals the body to make more red blood cells. As you know, more oxygen means the muscles can work longer before becoming anaerobic, increasing endurance.

Eighteen well-trained athletes ran a 3,000 meter time trial, and then used EPO for four weeks before running another time trial. The average improvement for the group was 6 percent when compared with their baseline—quite a jump in performance in just four weeks' time. What's more, the improvement endured even after they stopped taking EPO— another time trial four weeks after the drug use ended found they still were 3 percent faster than their first time trial.

That performance increase is in line with some of the other results we can infer from comparing performances from the Tour de France during the 1990s and 2000s with performances today. Generally, the fastest climbers in the tour are covering the biggest climbs, like the Alpe d'Huez, about 5 to 10 percent more slowly than they did a few years ago. Again, it could be years before we see clean athletes match those times.

Up on the Mountain

The first thing you notice when walking into Randy Wilber's office at the U.S. Olympic Training Center in Colorado Springs is the Wheaties boxes.

They're lined up precisely on a shelf above his desk, the orange logos framing photos of Olympians such as Apolo Anton Ohno, Joey Cheek, and Hunter Kemper. Each of them is signed with a variation of "Thank you, Randy." A poster on another wall is signed by members of the U.S. women's Olympic cycling team. They're all mementos from some of the hundreds of athletes that Wilber has helped during his twenty years working with the U.S. Olympic Committee. (My remark on another Wheaties box with Pittsburgh Pirates great Roberto Clemente on it leads to the discovery that Wilber and I grew up in the same small town in western Pennsylvania, a coincidence neither of us can quite process.) A set of bookshelves is filled with binders, years of research papers organized into clean sets labeled with titles like "Lactate Metabolism" and "Sports Nutrition." The precision of his office mirrors the workings of his mind. When you ask Wilber a question, there's a pause while he considers it, and then full paragraphs of nuanced, thoughtful answers flow with no pause or hesitation.

Wilber has spent much of his career focused on what he calls "environmental exercise physiology," the study of how the environment in which an event is taking place affects an athlete's performance. That's led to him doing a lot of work on heat and humidity, pollution (especially at the Beijing Olympics, an experience he calls the biggest challenge of his career), and most famously, altitude. Wilber is acknowledged as one of the planet's foremost experts on the effects of altitude and how it changes our bodies, our blood, and our abilities.

The basic science of altitude is straightforward: The higher we are, the less oxygen we get with every breath. It's not that there's less oxygen in the air; the percentage of oxygen in the earth's atmosphere is about 21 percent at all altitudes. But the atmospheric pressure decreases as you get higher, and lower atmospheric pressure means that there are fewer molecules for a given volume of air (those of you who remember your chemistry, this is Boyle's law). At seven thousand feet, the atmospheric pressure is 22 percent lower than at sea level, which means that you get 22 percent less oxygen per breath. That's why climbing a set of stairs can feel like a hard workout when you first arrive at altitude.

But our bodies are amazing machines for adaptation. In response to the lowered effective concentration of oxygen, we begin to produce EPO, signaling the body to create more red blood cells. By increasing the number of red blood cells that can carry oxygen to the muscles, the body does its best to ameliorate the effects of the lower amount of oxygen available. It's pretty amazing, if you stop and think about it.

Once your body adapts to altitude, it will have more red blood cells, which is a very good thing for an endurance athlete. Many scientists point to this altitude effect as part of the reason for the success of Kenyan and Ethiopian distance runners, the vast majority of whom live along the Gregory Rift, a ridge that runs through the two countries at high altitude. Runners like these live and train at high altitude, what's known in the altitude business as "live high, train high," or LHTH. But there's a potential problem with LHTH, according to Wilber. "If you take a runner like [American distance running star] Galen Rupp who has to do a workout where he's doing repeat fifteen hundred meters at race pace or faster, can he do that at altitude?" he says. "Probably not." The decreased oxygen makes it difficult to work at the same intensity as the athlete can at a lower altitude, even with increased red blood cells. "So not only is he not progressing; he's probably regressing," says Wilber.

The possible solution is "live high, train low," or LHTL. The athlete lives and does his low-intensity workouts at a high elevation, but for hard workouts, he goes to a lower altitude so that he can achieve a higher intensity. It's an attempt to marry the benefits of living at altitude with the benefits of training at a more normal elevation. Clever, right?

But it's kind of complicated. Wilber helped compile a set of guidelines for American athletes who wanted to do this sort of training. First, you need to be in the sweet spot when it comes to where you're living: too low and you don't get as much adaptation, but if you're too high, you can't recover or sleep as well as you need to. That sweet spot seems to be in the 6,500- to 8,000-foot range—perhaps not coincidentally, the elevation at which many Kenyan and Ethiopian runners live. This sweet spot was confirmed by a 2014 study that found that higher altitudes didn't lead to increased adaptation.

Next, you need to find a place to train low. The best practice recommendation from the USOC is that dropping down to four thousand feet or so will help most athletes in their quest for increased intensity. But the length of stay at this kind of altitude camp is very important. Wilber notes that living at altitude for two weeks is basically useless for an elite athlete in terms of real change to their red blood cell count. By three weeks, there's a 4 percent increase in red blood cell mass, but by four weeks, that increase nearly doubles to almost 8 percent. That's why the USOC recommends that athletes should plan on at least a four-week camp.

So: You need four weeks in a place where you can live at seven thousand or so feet and then somehow get to four thousand feet to do your most intense workouts. It's a daunting logistical challenge. Athletes in the United States are actually very lucky when it comes to LHTL. There aren't a lot of places in the world where you have the right combination of geography and transportation to allow you to follow the ideal regime, and we have several that have become popular with athletes. There's Flagstaff, Arizona, at 6,910 feet, which is a 31-mile drive from Sedona (4,326 feet). There's Mammoth, California, which sits at 7,880 feet and is 42 miles from Bishop, which is at 4,150 feet. And then there's Park City and Salt Lake City, Utah, where a 30-mile drive can take you from 7,000 feet to 4,327 feet. As Wilber says, they're places where it's easy to get "down the mountain" to train.

The mountain on which the Olympic Training Center (elevation 6,035 feet) sits is on a high plateau, making getting down a harder task. So instead, the USOC is bringing sea level to the mountain. In the center's new training building, Wilber and his team are constructing an environment that can be tailored to the specific needs of a given athlete. "We'll be able to take it to simulate up to twelve thousand feet, or to pressurize it all the way down to sea level for training low," says Wilber as he shows me a sketch of the chamber. "We'll also be able to apply heat and humidity so we can use it for acclimatization for places like Rio in the 2016 Olympics. It won't quite happen with the flip of a switch, but within an hour or two, we'll be able to exhaust one environment and apply a new one."

When the center is completed, there will be treadmills and stationary bikes, as well as cots to let athletes sleep at a specific elevation. For Wilber, who's dedicated twenty years to unraveling the effects of environment on athletes, it will be a way to dial up exactly the conditions that he needs to optimally tweak their performance.

But you could argue—and some people have—that a chamber like the one the USOC is building amounts to a kind of technological doping.

All's Fair in Love and . . . Blood Cells?

The most simplistic argument against drugs in sports takes just five words, and it's understandable by even a five-year-old: It's not fair to cheat.

But fairness can be a slippery concept. First of all, as you might have noticed, the world isn't fair. And sports aren't fair. Taking the concept of fairness to an absurdist extreme, if athletic competition were completely fair, every race would end in a tie. No one would have any advantage over anyone else, whether it was some sort of genetic proclivity toward a certain event or the economic resources that allow him to train more hours in a day than the competition.

That's the most radical view of fairness in sports. Now let's step back from the logical abyss, and look at another area in which it's hard to separate what's fair from what's not, what's cheating and what's just smart science and training. Let's talk about blood cells.

Scientists like Randy Wilber have spent a career unraveling the ways to use altitude to create more red blood cells in an athlete's body. No one has argued that living at altitude is somehow cheating and should be banned, even though it confers a performance enhancement.

Now, consider chambers like the one that the USOC is building in Colorado Springs, or altitude tents that simulate high-altitude conditions. Hypoxic tents, as they're called, are also legal for athletes, although it's not as clear-cut as natural altitude. After all, it's hard to ban anyone from living where they want to, but it's a little easier to think that spending thousands of dollars on a tent might be a little fishy. In 2006, the World Anti-Doping Agency actually examined the issue of artificial altitude, and was well along the path toward banning it. "There was a really interesting scientific and ethical debate," says Wilber. "You had people that argued that it was totally unethical, that it was the same thing as doping. And on the other hand, you had people saying that if you banned simulated altitude, you had to ban Powerade and Gatorade."

In the end, WADA decided to continue to allow the use of the tents. There is one exception: The tents are actually against the law in Italy. This caused an issue for the USOC, which had set up a hypoxic apartment for American athletes at the 2006 Winter Olympics in Turin, only to have to tear it out when the tents were outlawed in the country.

There are, of course, other ways to increase the amount of red blood cells in your body. You can get a blood transfusion. Blood doping, as it's called when the transfusion is done for performance-enhancing reasons, was studied as far back as the 1940s. The method really came to the forefront after the 1968 Olympics, when athletes from Kenya and Ethiopia, who lived and trained at high altitudes, dominated the endurance events at the games, which were held in Mexico City, at 7,300 feet above sea level. From that point until blood doping was banned in 1985, athletes from many countries used transfusions to boost their red blood cell counts.

Then, EPO came along. In 1987, the drug company Amgen patented rEPO, a version of the hormone that was created using the genes in the body that produce EPO, as a treatment for anemia and patients with kidney problems. It was a wonder drug for those patients—and for endurance athletes. Now, they could boost their red blood cell counts without having to deal with altitude or transfusions. The downside was that as their blood got red-cell rich, it got thicker, sludgy. Promising cyclists in their twenties started dropping dead of heart attacks, which isn't what you'd usually expect from such fit athletes. Until the first drug tests for EPO were introduced at the 2000 Olympics, its abuse was rampant. That testing, and now the use of the so-called "biological passport," which tracks an athlete's blood values over time and flags suspicious markers, have reduced but not eliminated the use of EPO.

EPO isn't legal in sports. Most of us view it, quite clearly, as cheating. But it's the same mechanism, same outcome as altitude: more red blood cells. It may be *more* more red blood cells, but that's a difference in magnitude, not kind.

Now it appears that some athletes are trying other ways to boost their production of EPO without injecting the hormone. There were reports around the 2014 Winter Olympics in Sochi that some Russian athletes were inhaling xenon gas in an attempt to increase EPO levels in their bodies. The gas appears to increase production of a protein called HIF-1 alpha, and increased levels of HIF-1 alpha lead to increased EPO levels and more red blood cells. If the Russian athletes were doing so, it would have been completely legal under the drug-testing rules then in force, but WADA banned xenon use in May 2014—even though there is no test to detect it.

Let's take the fairness argument to another extreme. If all of these things—from living in Boulder to shooting yourself up with EPO—have the same effect, why not make them all legal? Why not let athletes and scientists do whatever they would like to do in the pursuit of better performances?

This can be a seductive intellectual position; I've made this argument myself in the past. One way you can talk yourself into this stance is by embracing the idea of a level playing field. Many athletes who have admitted to doping during their careers say that they didn't want to, but because they suspected that their competition was doing so, they had to do it to keep up, to level the playing field.

Here's the thing. People having access to and using the same drugs doesn't level the playing field. Because just like everything else in sports, the responses to doping aren't fair, either. They're individual. Tyler Hamilton and Daniel Coyle, in their detailed exposé of doping in cycling, *The Secret Race*, explain the issue:

> For example: Hamilton's natural hematocrit is typically 42.
> Taking enough EPO to get to 50 [the limit at the time] means

he could raise his hematocrit 8 points, an increase of 19 percent. In other words, Hamilton could add 19 percent more oxygen-carrying red blood cells—a huge increase in power—and still test under the hematocrit limit. Now let's consider a different rider who has a natural hematocrit of 48. Under the 50-percent rule, that rider could only take enough EPO to add 2 points, or 4 percent more red blood cells—a power increase one-fourth of Hamilton's. That might be one of the reasons Hamilton's performance increased so rapidly when he started taking EPO. Also, studies show that some people respond more to EPO than others; in addition, some people respond more than others to the increased training enabled by EPO. Then you have the fact that EPO shifts the performance limits from the body's central physiology (how much the heart pumps) to the peripheral physiology (how fast the enzymes in the muscles can absorb oxygen). Bottom line: EPO and other drugs don't level the physiological playing field; they just shift it to new areas and distort it. As Dr. Michael Ashenden puts it, "The winner in a doped race is not the one who trained the hardest, but the one who trained the hardest and whose physiology responded best to the drugs."

The reality is that the playing field can never be level. Nor do we actually want it to be. We watch sports for winners and losers, for transcendent performances and heartbreaking failures. A truly level playing field would remove all of the things that make competition compelling to us.

It's clear how steroids and drugs like EPO can affect performance, but even athletes in sports such as archery and shooting have their performance drug of choice: beta-blockers. Usually used to treat cardiac arrhythmia, these drugs are ideal for precision sports. They reduce heart rate and suppress the flow of adrenaline, making for better aim and accuracy. (Beta-blockers are also highly valued among musicians for their anxiety-reducing properties.)

All of these substances violate the World Anti-Doping Agency code, which states that a substance must meet two of three criteria to be prohibited: (1) it enhances, or has the potential to enhance, performance; (2) it's an "actual or potential health risk" to the athlete taking it; and (3) it's contrary to "the spirit of sport" (the fairness argument).

Another often-used argument to eliminate the rules against doping is around enforcement. Since we can't catch every athlete who dopes, the suggestion goes, we should allow any athlete who wants to dope to go ahead. Again, this is an essentially absurdist position. We don't catch everyone who robs a house or kills someone—people do get away with murder—but our inability to perfectly enforce those laws doesn't lead us to think that we should do away with the laws altogether.

It's easier to make this argument in sports because of the seemingly arbitrary lines drawn between what's legal and what's not, like in the preceding red blood cell example. It's important to understand that *all* of sports is arbitrary. There's no survival imperative that requires that human beings play baseball, or that the bases be ninety feet apart, or that the bat be round or made of wood or any of the other hundreds of rules that make up the game. A sport is nothing more than a set of arbitrary rules that we all agree to.

The best argument that I know of for why we should ban doping comes at the problem from the opposite of the usual angle. Don't think of rules against doping as a way of punishing those who choose to violate them. Think of them as an attempt to protect the interests of those athletes who *don't* take drugs or cheat. If I want to compete at the highest levels, I shouldn't have to take drugs that have profoundly dangerous side effects. I shouldn't have to endanger myself in that way. Rules about doping aren't just about catching the cheats—they're about protecting the ability of athletes who don't want to assume the risks of doping to win.

Mice and Men

Athletes and coaches have long talked about "muscle memory" when it comes to strength training, as a way to explain a common observation: It seems to be easier for someone who has done a lot of strength training in the past to regain his or her strength when compared to someone starting from scratch. If I'm an athlete who used to be able to deadlift five hundred pounds, I can regain that ability much more quickly than someone who has never weight-trained before.

The working theory had long been that this was a motor skill—that having lifted before, the body retained some "memory" of how to do so. But research from the University of Oslo points to a very different reason for that muscle memory, one that could profoundly change our view of performance-enhancing drugs as well.

The research focused on a quirk of muscle cells: Unlike most other cells, they have multiple nuclei. When muscle mass increases (from weight training, for instance), the number of nuclei also increases. But there are also chemical ways to increase muscle mass, like taking steroids. When the researchers treated female mice with testosterone for fourteen days, they saw a 77 percent increase in muscle mass and a 66 percent increase in nuclei in the muscles. Then, they stopped the treatments, and in three weeks, the juiced mice were the same size as their nontreated, control counterparts.

But here's where it gets interesting: The number of nuclei remained elevated in those mice who had taken the steroids. And not just for a little while, but for three months, which is 10 percent of a mouse's life span. When these mice were put on an exercise program, their muscle mass increased very quickly—by 31 percent in just six days—while the controls didn't see any significant change.

If human muscle operates in the same way (and admittedly, this is a huge if—this sort of result in mice is no guarantee that it operates the same way in humans), it points out a huge hole in our current anti-doping system. The bans for doping offenses have traditionally been two years, although WADA recently announced that the ban for a first offense would be increased to four years beginning in 2015. But if even short-term doping could change the makeup of muscle in the way this study suggests, the effects of doping may last much, much longer than even that four-year span. The mice's nuclei gains lasted for more than 10 percent of their life span, which suggests that humans might experience benefits for many years.

This is an example of an epigenetic change, an environmental factor that causes an organism's genome to express itself in a different way. The results of the Oslo study strongly suggest that doping can lead to these sorts of changes, but there are other studies that show that the use of drugs such as growth hormone and insulin growth factor also lead to epigenetic effects. In a review of epigenetics in sports, scientists from the Johannes Gutenberg University of Mainz, in Germany, warn, "Even though the direct effect of doping substances on the body is mostly transient, epigenetic consequences might be persistent."

The road to clean sports isn't easy, and it's a journey that's never going to be complete. But I really do think we're making progress, slow as it may be. When you hear cyclists like Tyler Hamilton or Floyd Landis talk about the toll that doping and the deception it required took on them, it's moving. Even Lance Armstrong, who perpetrated the biggest scheme of all his peers, has shown emotion in a way that he never did in the past, cracks in the facade that had seemingly been so perfect for so long. There are always going to be people who are willing to do whatever it takes to win, even if it breaks laws or puts them at risk. But through education and science, the sports world has started to turn the tide in the fight against doping. And that's how we'll hopefully limit drug use to the few, rather than tacitly accepting it for the many.

12

THE LIMITS OF PERFORMANCE

Is the Curve of Human Accomplishment Flattening Out?

In the February 1935 edition of *Amateur Athlete*, the official magazine of the Amateur Athletic Union, there's a story outlining the scientific findings of Brutus Hamilton, the head track coach at the University of California. According to Hamilton, "the mathematical laws dealing with physiological compensation" set certain limits on human performance in track and field that could never be overcome.

Hamilton claimed that as he wrote, there were already two "perfect" world records that would never be improved. He was referring to the 400 meter record of 46.2 seconds and the shot put record of 57 feet 1 inch. Today, the 400 meter record is 43.18 seconds; the shot put record is 75 feet 10¼ inches.

I don't bring this up to mock Hamilton, although that's an easy knee-jerk reaction to outdated predictions of the future. I bring it up as a cautionary tale. Almost every time someone has claimed that we've reached the outer boundaries of our abilities as a species, the statement has usually been proven false in a distressingly short time. It's one of those quirks of science and human perception: If you predict that humans will keep progressing, it's hard to be proven wrong, per se. You can just argue that the progress hasn't happened *yet*. But if you argue that we've reached the limits of performance, all it takes is a new world record the next day to prove you wrong.

But there's an irresistible urge to try to understand how close we may or may not be to the outer boundaries of our abilities. Take something as elemental as running. Human beings have been running for the entirety of our existence as a species. Evolutionary biologists such as Harvard's Daniel Lieberman believe it was actually our ability to run for long periods of time that enabled humans to survive—we hunted successfully by simply pursuing our prey to exhaustion. The fact that we sweat rather than pant to regulate our body temperature; the way our legs store energy while we run, almost like springs; even the size of our rear ends—all of these are thought to be adaptations that enable us to move efficiently and run long distances without reaching exhaustion. In the words Christopher McDougall used for the title of his book, in many ways it appears that humans were "born to run."

Given that long evolutionary history, the improvements we've made as distance runners over the past century or so are even more remarkable. On the afternoon of April 10, 1896, seventeen men from five countries set out to run the first Olympic marathon. Doctors warned that such an exertion might be extremely dangerous, and in fact only eight of the competitors successfully finished the contest; a Greek horse groom named Spyridon Louis won the race in 2:58:50. In Beijing 112 years later, Kenya's Samuel Wanjiru won the marathon on a sweltering day with a time of 2:06:32. After not much more than a century, Wanjiru ran his race—more than 2 kilometers longer than Louis's, by the way—29 percent faster.

In sport after sport, we can see a similar progression in our physical abilities at the elite level of competition. In swimming, the men's 100 meter freestyle record has improved 29 percent since 1905. The men's shot put world record is 49 percent farther now than in 1909. For women, the gains are even greater—some women's field-event records have improved by more than 100 percent.

But after a century of massive accomplishment, the pace of improvement has slowed dramatically in the past twenty years. From 1905 to 1988, the men's 100 meter freestyle swimming record dropped an average of 0.32 percent a year. In the 24 years since 1988, it has dropped just 0.13 percent a year.

You see the same thing in many sporting events. The curve of our improvement is flattening out. A mathematician sees a curve that's flattening and draws a simple conclusion: We're reaching some sort of limit.

Those curves are what caught the eye of Geoffroy Berthelot, a researcher at the French sports research institute known as IRMES. Berthelot and his coauthors looked at the world records in five sports—cycling, speedskating, swimming, track and field, and weightlifting—beginning in 1896, at the dawn of the modern Olympic era. At the start of that time period, the improvements came fast, as we've noted. But Berthelot found that over the past two decades, the improvement has slowed so dramatically that he calculates that half of all world records are unlikely to improve by more than 0.05 percent before 2027. And that goes for every single event he looked at: "In all measurable Olympic contests from five different disciplines, involving either aerobic (10000 m skating) or anaerobic (weight lifting) metabolic pathways, leg muscles mainly (cycling) or all muscles (decathlon), lasting seconds (shots) or hours (50 km walk), either in men or women, small (Fly weight) or tall athletes (100 m free style), individual or collective events (relays), all progression curves follow the same pattern, supporting the universality of the model."

So is that it? Are we humans so near those boundaries that our kids' kids won't see new world records? I don't think so. What I do think is that it's going to be much, much harder to break those records.

All the science and technology we've talked about in this book is aimed at helping to push those records forward, as scientists help athletes past the former bounds of their performance to a new level of achievement. But at a certain point, logic suggests, there's no further that they can go.

In an evolutionary sense, our species has been focused on sports as opposed to survival for a laughably short period of time, a proverbial split second. If you think of sports as an experiment, it's one we've only really been engaged in for a little more than a century. Sure, there are some sports and games that have been operating at a high level longer—the modern rules of soccer were drafted in 1863, baseball's National League was founded in 1876, and the first cricket test match was played in 1877. However, even at the time Hamilton was making his predictions in 1935, only thirty-seven nations had competed in the most recent Olympics, with just over 1,300 athletes (only 126 of whom were women). In 2012, there were 204 countries represented at the London games, with 10,568 athletes, 4,676 of them women.

In the past century we've brought the opportunity to compete to millions more potential athletes around the world, expanding the talent pool exponentially. This gives us more variation and a better chance to find truly unique individuals who are particularly suited to athletic greatness. When you use population to model the increase in athletic performance, it turns out that it's a pretty good predictor of our progress—Scott Berry, writing in the magazine of the American Statistical Association, calculated that in several sports, more than 80 percent of improvement over time has come simply from the growth in the world's population.

But even with all those new athletes, even with all of the advances we've made in our understanding of the human body, there are still places in the sports world where progress hasn't just stalled, but has gone backward. In those events, the dream of constant progress has gone from aspiration to mockery.

For instance, what the hell is going on with the long jump?

Takeoffs and Landings

The evening of August 4, 2012, was later dubbed Super Saturday by the British papers. Within the space of an hour, three British athletes clinched gold medals in track and field at the London Olympics. Jessica Ennis started things off by winning the heptathlon, and Mo Farah ended the evening by winning a thrilling race to take the gold in the 10,000 meters. In between, there was long jumper Greg Rutherford.

Rutherford, who until 2012 had never been much of a factor in a major international competition, wasn't among the favorites in the event. Nonetheless, he outjumped the rest of the field, clinching the gold. His winning jump was just 8.31 meters, the shortest gold-medal-winning jump since 1972. Admittedly, that's still more than 27 feet 3 inches, a long way to fly through the air. But in the history of the event's best results, Rutherford's leap was number *574*.

Compare that with Carl Lewis, who won gold in the long jump in four consecutive Olympics. Lewis jumped farther than 8.31 meters forty-seven times over the course of his career.

How different has the pace of improvement been between the long jump and other track and field events? Jesse Owens was one of the greatest track athletes of all time, but his gold-medal-winning time in 1936 of 10.3 seconds for the 100 meters has long since been eclipsed; he wouldn't have made it to the quarterfinals in London with that time. Owens's long jump of 8.06 meters from 1936? It would have placed him seventh at the 2012 Olympics. It's not just at the very top of the heap that the standard has fallen. Of the top 107 long jumps of all time, only twenty-three of them have happened since 2000. Clearly, something is going on in this event that keeps it from following the pattern of improvement we see in almost every other sport. Just look at the results in meters of the long jump from every Olympics and World Championships.

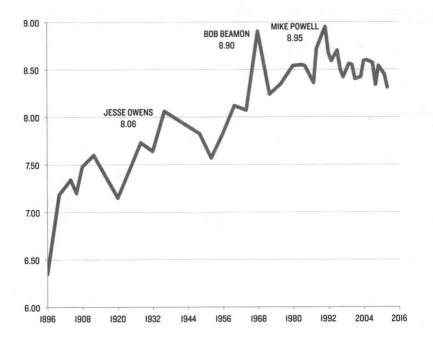

Carl Lewis seems to take the decline of long jumping personally. He was in London, and talked to the gold medalist after the event. "I saw Rutherford and I congratulated him because I thought he jumped well . . . for him," says Lewis. "But I was very clear that it was an embarrassment and that he needed to jump farther. It's just an embarrassment that those performances win Olympic medals. These marks people are jumping, it's what people were jumping when I came into the sport in the 1970s. They're just winning medals by default."

Why is this the case?

It's not that we're running more slowly, that's for sure. Because, after all, the first variable needed for a, well, *long* long jump is raw speed. The faster an athlete is traveling when he hit the board at the end of the runway, the longer we'd expect him to jump. It's estimated that Carl Lewis reached a velocity of more than 11 meters per second at takeoff—remember, he was the world record holder in the 100 meter sprint as well.

That record is long gone, of course. Usain Bolt has demolished the 100 meter record over the past several years, and races like the 2012 Olympic final featured the fastest field in sprint history. During Bolt's world record run of 9.58 in 2009, he reached a velocity of 12.27 meters per second.

But being fast isn't enough. "I hear people say that you just run down to the board and jump, and that if you have speed, you'll be good," says Lewis. "That's just like saying if you can throw the ball fast, you're a good pitcher." The other key aspect of the long jump is how the athlete translates his horizontal velocity into vertical velocity, how he goes from running to jumping through the air. The more efficiently you translate between the two, the farther you'll fly, but it's not easy to do well. ("To really jump far, you have to come off the board feeling like you're going to fall on your face," says Lewis.) Mike Powell, for instance, couldn't beat Lewis in a footrace, but the vertical portion of his jump was stronger than Lewis's. That's how Powell ended up with the world record.

Using some fancy physics modeling, we can predict the distance that a long jumper could attain at a given speed. If you plug Bolt's top speed into the model, you get an otherworldly number: 10.50 meters. Even if you reduce the efficiency of Bolt's takeoff by half of what the model predicts, you still get an estimate of 9.46 meters, which would shatter the world record.

There's just one problem. Bolt doesn't compete in the long jump. Unlike Lewis or Jesse Owens, Bolt has never taken his world record speed to the long jump runway. Long jumping is hard, and all of the takeoffs and landings create an increased chance of injury. With an income estimated by *Forbes* at more than $20 million a year, there's basically no incentive for Bolt to risk injury to try a new event. That seems to be a common theme—none of the top sprinters in the world compete in the long jump, leaving the event to specialists who don't have quite the same unearthly speed.

And even if there were a jumper in the world who was able to challenge Powell's world record, he might have a hard time of it. Part of what enabled Powell and Lewis to reach such lengths in the event is that they had one another as competition.

Their rivalry culminated in the greatest night of long jumping ever seen. Going into the 1991 World Championships in Tokyo, Lewis had won sixty-five consecutive meets over the course of ten years, and was regarded as the greatest jumper in the history of the sport. At the start of the finals, he jumped 8.68 meters to take the lead, and then stretched his lead with a wind-aided 8.83 meter jump.

On his fourth attempt, Lewis flew 8.91 meters, farther than Bob Beamon's legendary world record, although a strong tailwind meant that it wouldn't be considered a legal record jump. All Powell did in response was jump 8.95—soaring past Lewis and, since the wind had died down, setting the new record. "I expected him to jump well, but I didn't think he'd do that," says Lewis. "I'll just be honest. It took me a minute to get it together."

Lewis, after watching Powell celebrate, wanted to get right back into the competition. "I wasn't next in the order, but I took off my warm-up clothes and got on the runway," he says. "They were like, 'It's not your turn.' But I was ready to go; I had those two more jumps. I was ninety-nine percent sure that one of those last two jumps was going to be longer and I'd break the world record."

It wasn't enough. In his attempts to catch Powell, Lewis traveled 8.87 meters and 8.84 meters in his last two jumps. Before that day in Tokyo, only two men had ever jumped more than twenty-nine feet; Lewis and Powell did it in less than an hour. They made three of the seven longest jumps in the history of the event, and no one has come close since.

"Mike Powell wasn't as fast as I was, but he had the intangibles," says Lewis. "The way he was aggressive, how he came off the board with no fear. Even fast people, if they don't have that, it doesn't matter. We were fierce competitors, and we're good friends now. And I respected him so much. It took him ten years to chase me down, and I really respected that."

That duel, the ability for two athletes to one-up each other past their previous best, was reminiscent of what happened when Roger Bannister broke the four-minute barrier for the mile. Australian John Landy had been trying for years to crack four minutes without success, and then just six weeks after Bannister's historic run, Landy broke Bannister's world record. There's something powerful about seeing a competitor do something that's been out of reach for you. First, the competitive instinct is very strong and highly motivating. But it also shows you that it's possible, and that's important. The four-minute mile went from a "barrier" to reality when Bannister ran it. (That competitive drive is part of what make sports so compelling, after all.)

Lewis points out that many of the more technical events in track are suffering from the same stagnation. "Look at the pole vault; look at the high jump," he says. "The high jump is going nowhere; the triple jump is going nowhere. None of these technical events are getting better." Lewis points to a lack of coaching expertise as a major factor in all of these events, saying that too many coaches are former athletes trying to extend their careers in the sport. "Just because I could jump far and run fast doesn't mean I'm qualified to coach," he says. "As I learn to coach, I'm asking people to mentor me and help me become better."

It's almost a perfect storm—a highly technical event, a lack of coaches, the very best athletes not participating in the event, a gradual decrease in the expectation of performance around the world. There's no single factor that you can point to that explains the drop in long jump performance, but rather the interaction of a lot of small factors. The way to get the event back on track is to focus on each of these issues, and try to move all of them forward.

That's what Lewis is doing. He's looking to put his effort where his criticism is, as he works with University of Houston sprinter Cameron Burrell (a son of former 100 meter world record holder Leroy Burrell) to turn him into a great long jumper.

"He's got the temperament, the aggressiveness, the speed," says Lewis. "If Cameron is successful, then I'm right. And if he's not, I'm wrong."

Perfect Pitching

The San Diego Padres led the Cincinnati Reds 4-3 in the bottom of the eighth inning on September 24, 2010. Both teams were fighting for a playoff spot as the season wound down, so the series was a big one for each club. With one out in the inning, Tony Gwynn Jr., the son of the legendary Padre hitter, dug into the batter's box to face rookie Aroldis Chapman. Chapman had defected from Cuba the year before, but it was just his second month in the major leagues.

The tall, thin left-hander reared back and threw his first pitch to Gwynn, a 103 mph fastball for a called strike. Next pitch, a 104 mph ball up and in to Gwynn. Two fastballs were fouled off, one at 103, one at 104.

So the count was 1-2 as Chapman threw his fifth pitch of the at bat. It missed the strike zone, just off the inside corner of the plate. But that's not what made it noteworthy. It's what the radar gun read: 105 mph. To be precise, the pitch was actually measured by baseball's Pitchf/x tracking system at 105.1 miles an hour, making it the fastest pitch ever thrown.

Or not.

Maybe.

It's definitely the fastest pitch ever tracked by the Pitchf/x system, which debuted in 2006 and was installed in all thirty major league ball-parks by 2007. But baseball history is riddled with claimants of the title of fastest pitcher ever. Nolan Ryan dominated the game for decades with his fastball, and was the longtime holder of the Guinness World Record for fastest pitch. Negro league legend Satchel Paige is said to have thrown at otherworldly speeds; Hall of Famer Bob Feller was known as "Rapid Rob-ert" for his velocity. Going back to the 1920s, Walter Johnson is still cited by many as one of the hardest throwers the game has ever seen. Those older pitchers were never measured with a radar gun, although there were efforts made to time their pitches in races with a motorcycle or at a weapons proving range.

Then there's the case of Steve Dalkowski. Dalkowski was an average-sized man, 5 feet 11 inches, 175 pounds. But he had one otherworldly ability: He could throw a baseball insanely hard. Pat Jordan, writing in *Sports Illustrated*, told the story of the day the legendary Ted Williams stepped into the batter's box during batting practice to see just how hard Dalkowski threw. Dalkowski fired a fastball past him, who couldn't quite believe its speed:

> The catcher held the ball for a few seconds a few inches under Williams' chin. Williams looked back at it, then at Dalkowski, squinting at him from the mound, and then he dropped his bat and stepped out of the cage. The writers immediately asked Williams how fast Steve Dalkowski really was. Williams, whose eyes were said to be so sharp that he could count the stitches on a baseball as it rotated toward the plate, told them he had not seen the pitch, that Steve Dalkowski was the fastest pitcher he ever faced and that he would be damned if he would ever face him again if he could help it.

(Years later, Dalkowski told Jordan that this story was apocryphal, and that he had never pitched to Williams. As Jordan said to me, "It was a story that we all heard in the minors in 1959 or 1960. If it wasn't true, it should have been.")

Dalkowski pitched nine seasons in the minor leagues, making it as high as Triple-A but never cracking the major league roster. The problem wasn't his velocity or his ability to strike out hitters: In the 995 innings he pitched in his professional career, he struck out 1,396 batters. That's 12.63 strikeouts per nine innings, a breathtaking number. Unfortunately, Dalkowski couldn't get the ball over the plate. In those same 995 innings, he walked 1,354 batters. His velocity and wildness led to games like the one he pitched in 1957 for the Kingsport Orioles. He struck out twenty-four opposing batters . . . and lost, 8-4. Because he'd also walked eighteen, hit four batters, and thrown six wild pitches.

Dalkowski's career has grown in legend in the years since he retired in 1966—he actually served as the model for hard-throwing pitcher Nuke LaLoosh in the classic baseball movie *Bull Durham*. Baseball loves these sort of characters, and it's increasingly difficult to separate fact from fiction at this point as Jordan's story shows. But plenty of people in the game believe that Dalkowski might have thrown harder than anyone else ever has.

There's good evidence that humans actually evolved to become powerful throwers. The research comes from the lab of Daniel Lieberman. One of Lieberman's students, Neil Roach, was the lead author of a paper in *Nature* that argues that "selection for throwing as a means to hunt probably had an important role in the evolution of the genus *Homo*."

Our nearest biological relatives, the chimpanzees, don't throw many things (and when they do, it's often feces. Which is a whole different problem.). It's very rare that chimps will throw anything in an overhand motion, and when they do, they use more of a full-arm slinging motion, like a cricket bowler rather than a baseball pitcher.

Launching a projectile with a straight arm like in cricket is much less efficient or powerful than throwing. The fastest ball ever bowled in cricket came from Pakistani fast bowler Shoaib Akhtar, whose ball was clocked at 100.23 mph in the 2003 World Cup. That's obviously very quick indeed, but the cricket bowler has the advantage of a running start before he delivers the ball to help overcome the inefficiency of the bowling motion. Akhtar bowled very hard, but there are very few cricket bowlers who've hit the midnineties on a radar gun in cricket history, while baseball pitchers routinely throw that hard.

It turns out that chimpanzees throw like cricket bowlers because they don't have the anatomy to throw like humans. As we evolved separately from chimps, we developed shoulders that allowed us to be much better at throwing things than our ape cousins. When we throw like a baseball pitcher, we cock our arms back, stretching the connective tissues that run across our shoulders. "Those tendons and ligaments get loaded up like the elastic bands on a slingshot, and late in the throw they release that energy rapidly and forcefully to rotate the upper arm with extraordinary speed and force," said Roach in a Harvard press release about his study.

Those speeds and forces really are extraordinary. We think of contact sports being violent—and they are—but the most violent movement in all of sports is the acceleration of a pitcher's arm as he throws a ball. In fact, the rotation of a pitcher's arm is the fastest motion that the human body can produce. "The rotation of the humerus can reach up to nine thousand degrees per second, which generates an incredible amount of energy, causing you to rapidly extend your elbow, producing a very fast throw," Roach said.

That's a tremendous amount of stress to place on the body, and it habitually points out the weakest link in what biomechanists call the "kinetic chain." That weak link is a ligament on the inside of the elbow that connects the upper arm to the ulna, one of the two bones in the lower arm. It's called the ulnar collateral ligament, and while you might not know the ligament's name, many sports fans will know the name of the operation that's performed to repair a rupture of the ligament: Tommy John surgery. It's named after the first athlete to undergo the operation, former Dodger and Yankees ace Tommy John, who had his career extended for thirteen seasons after getting his elbow injury fixed.

Until that point, an ulnar collateral ligament rupture basically ended the career of a pitcher. Since that pioneering surgery by orthopedist Frank

Jobe in 1974, the operation has become commonplace. Baseball injury expert Will Carroll reported in 2013 that 124 pitchers on major league rosters had undergone the procedure—fully one-third of all pitchers at the game's top level.

So, in baseball, we've got a game in which there's at least anecdotal evidence that pitchers have been throwing at the same level of top speeds for a century. We've got a clear pattern of injury, and a culprit in the ulnar collateral ligament. All of these things could lead you to a pretty clear conclusion: Baseball pitchers have reached the limit of what they will ever do with our current anatomy.

That's the opinion of Glenn Fleisig, a biomechanist at the American Sports Medicine Institute and one of the world's foremost experts on the science of pitching. Fleisig has done a simple but telling test using cadaver ligaments. You can take one of those ligaments and see how much force it can handle before it breaks. When you do that experiment, you find out that it's just about the same force created when a pitcher throws at 100 mph.

"There may be an outlier, one exception here or there," Fleisig told the *New York Times*. "But for major league baseball pitchers as a group of elites, the top isn't going to go up anymore. With better conditioning and nutrition and mechanics, more pitchers will be toward that top, throwing at 95 or 100. But the top has topped out."

We can strengthen our muscles to provide more force, and we can work on our mechanics to use our muscles more efficiently. But our tendons and ligaments can't really be strengthened in the same way—they can handle what they can handle, and there's not much we can do about it. Some athletes might have stronger tendons, and that might enable them to get a little more on their fastball without blowing out their elbow. As a species, we might have evolved to throw hard, but we didn't evolve to throw hard as often as pitchers now do. As Roach writes in his paper:

> Paleolithic hunters almost certainly threw less frequently than
> modern athletes, who often deliver more than 100 high-speed

throws over the course of a few hours. Unfortunately, the ligaments and tendons in the human shoulder and elbow are not well adapted to withstanding such repeated stretching from the high torques generated by throwing, and frequently suffer from laxity and tearing. Although humans' unique ability to power high-speed throws using elastic energy may have been critical in enabling early hunting, repeated overuse of this motion can result in serious injuries in modern throwers.

In that at bat against Aroldis Chapman, Tony Gwynn eventually struck out, taking a called third strike on a fastball, naturally. It clocked 102 mph. After the game, Gwynn told reporters about Chapman's record pitch, "I didn't see the ball until it was behind me." The good news for batters is that science suggests they won't have to adjust to anything much faster in the future.

Humans and Horses

Let's approach the question a different way. There are other sporting events in which the pursuit of speed has been even more controlled than human track and field. We humans have been trying to create the fastest horses and dogs possible for generations. We take a horse, see if it's fast, and if it is, we'll breed it with other fast horses, trying to squeeze every bit of genetic advantage from the population. If human sport is the unplanned experiment one researcher described it as, horse and dog racing is—very clearly—a well-organized genetics program.

Horse racing's Triple Crown annually pits the very best three-year-old Thoroughbreds against one another. The three races—the Kentucky Derby, the Preakness Stakes, and the Belmont Stakes—have all been contested since the 1860s and '70s. In all those years, only eleven horses have swept the three races to take horse racing's ultimate prize, and none since Affirmed in 1978.

There's something else that hasn't happened since the 1970s in the Triple Crown races: The horses haven't gotten any faster. Performance in the Kentucky Derby reached a statistical plateau in 1949, while the Preakness and Belmont reached plateaus in 1971 and 1973, respectively. It was in 1973, incidentally, that Secretariat, maybe the greatest three-year-old ever, swept the Triple Crown and set the record time in each of the races. Those records all still stand, forty years later. Forty years of breeding and training, and we don't have any horses that can run any faster.

The same phenomena can be shown in greyhound racing. In the three largest English greyhound races, all of which started in 1927, the speeds reached a plateau between 1966 and 1971.

These insights come from a paper written by Mark Denny, a professor at Stanford whose usual field of study is marine science biomechanics. Denny's also a recreational runner, and he was drawn to the question of how fast humans might be able to run in the future. To try and find out, he looked at those horse racing and dog racing results to see if they revealed anything about the progression in those sports. His conclusion?

> Despite intensive programs to breed faster thoroughbreds and greyhounds, despite increasing populations from which to choose exceptional individuals, and despite the use of any undetected performance-enhancing drugs, race speeds in these animals have not increased in the last 40–60 years.

What Denny is saying is that we shouldn't expect horses or dogs to run any faster, given the statistical analysis of their race times and the size of the population of animals available. At the most, he predicts that no more than a 1 percent increase in Thoroughbred performance is possible.

Where things get more complicated is when you apply that same analysis to human races. Denny predicts that we'll see more improvement than we will in horses and dogs, but not much. We might see up to a 5 percent improvement in track world records, depending on the event, but that's it. "One of the fun things to me was that you look at how much better you can get at each one of the distances," says Denny, "and it was more or less the same, from the hundred meters to the marathon. But the mechanics of what's limiting us must be totally different in those events. To have the projections be that we can get two or three percent better in each one is really interesting."

Based on his analysis, Paula Radcliffe's marathon world record is the closest that a human being has ever come to reaching our absolute maximum as a species; her time of 2:15:25 at the 2003 London Marathon is just 0.33 percent slower than Denny's predicted absolute record of 2:14:58.

(An interesting aside: Many of the same nutritional techniques that humans can use to increase performance are effective in the animal world. For instance, we've talked about cellular buffering, using something like sodium bicarbonate, and its effect on human performance. That same practice is banned in horse and dog racing.)

Each of the three athletic worlds that Denny examined reacted differently to his conclusions. "The horse people who put billions of dollars into breeding horses—the thought that the horses hadn't gotten faster didn't sit well with them," he says. "They said the horses are getting faster, but the tracks were getting slower. I heard nothing from the dog people, I think in part because of the animal rights issues. On the human side, people get really upset at the idea that there are limits. It strikes a bad chord with them."

You hear that skepticism when you talk to athletes, who are certain that they're not near any limits of performance. "Part of why I enjoy doing track is that I know there's room to improve in everything I've done," says Ashton Eaton, the world record holder in the decathlon. "I still know I can run faster; I still know I can jump higher. As far as we can see, we only keep getting better. We keep going so we can see the limits."

Denny's predictions, of course, carry the echoes of the predictions made by Brutus Hamilton at the start of this chapter. That's the danger of any prediction that says we're near our limits—somehow, we always seem to surpass them. If that happens, if Denny and Berthelot turn out to be wrong, Denny, for one, says he won't mind.

"Every time Usain Bolt gets on the track to run a hundred meters, I cringe a little bit to think he might run 9.48 [the fastest 100 that Denny predicted]. But it makes it interesting to watch those races now. I feel like I have a little skin in the game.

"On the other hand, I'd love to have someone exceed my predictions because it would just be really cool to have someone go that fast. Those times just seem inhuman to me. If somebody did it, I wouldn't be sad that my prediction went down in flames."

A Different Question

The first thing you notice when you enter the nondescript redbrick warehouse in Santa Monica, California, is a massive skateboard ramp that undulates across the entire space, the wood perforated here and there so that building inspectors consider it an art installation rather than an extreme sports playground. You make your way to the reception desk, but instead of being offered water, you're offered a can of the drink that made the existence of this office possible, the drink that funds hundreds of athletes and, now, a cadre of sports scientists as well. You're offered a Red Bull.

Over the past thirty years, Red Bull has grown from its roots as a Thai "tonic" called Krating Daeng into a $6.7 billion company. Its founder, Austrian businessman Dietrich Mateschitz, is worth an estimated $9.2 billion. An integral part of the brand's strategy from almost the beginning has been its association with sports, especially extreme sports. In 2012, Red Bull spent tens of millions of dollars to fund Felix Baumgartner's supersonic parachute jump, and it sponsors athletes in skiing, snowboarding, skateboarding, and more. In the 2014 Sochi Winter Olympics, if Red Bull were its own country, it would have finished sixth in the medal standings, just behind Germany and just ahead of Austria, as the company's sponsored athletes took home eighteen medals. Beyond those sports, Red Bull runs not one but two Formula 1 auto racing teams and owns three soccer teams around the world.

"We have more than six hundred athletes that we sponsor in over one hundred sports," said Andy Walshe, Red Bull's director of high performance, in a glass-walled conference room overlooking the skate ramp and the rest of Red Bull's North American headquarters. "You can't get that talent pool at a national sports governing body or anywhere else."

This talent pool—and Mateschitz's billions—has allowed Red Bull to establish an entire high-performance sports program overseen by Walshe, who is Australian and a biomechanist by training. Joining him in the conference room are other members of Red Bull's scientific circle. Per Lundstam, a Swedish trainer, came from the U.S. Ski Team, where he worked with Walshe. There's South African physiologist Holden MacRae, a professor of sports medicine at Pepperdine University. Clint Friesen works with motocross riders at a training center in Georgia to increase their physical capacity.

The group is huddled around a huge table in the middle of the conference room. Strewn across it are dozens of charts and graphs, brightly colored lines tracing arcs representing respiratory flows, or hemoglobin levels, or power outputs in cycling tests. Thick dossiers with endless columns of numbers summarize blood chemistry, sleep quality, and nutritional intake. The goal of the gathering? To try and make sense of all the data arrayed before them and turn it into actionable information for some of Red Bull's sponsored athletes.

The data was collected throughout a five-day intensive camp that the group dubbed Project Endurance. The goal of the camp, held at Death Valley and White Mountain, in the Sierra Nevadas, was both simple and ambitious: Gather a group of great athletes and a dozen or so top scientists, do a bunch of testing and trials at both sea level and altitude, and see what you can learn about what limits our endurance. Is it our hearts? Is it our lungs? Is it our minds? To test the effects of altitude change, they took a portable lab to the two locations, going from below sea level to nine thousand feet in a day, and ran the athletes through a battery of physiological tests and trials. During these efforts, the athletes were hooked up to heart rate monitors and glucose meters. They used a device called Physio-Flow that functions sort of like an EKG, to show cardiac output. They wore NIRS devices—near infrared spectroscopy—which use light to measure the level of hemoglobin in a particular muscle, and a device called Moxy, which gives information about blood oxygen levels and blood flow to muscles. Sensors also captured the level of oxygen in the brain and measured the gases athletes breathed in and out. At various points, each athlete at the camp was wired up with more than fifty thousand dollars' worth of sensors, wires, tubes, and electrodes.

It's this level of attention and resources applied to great athletes that holds the key to progress in our performance. It's the great joy and fear about working on the cutting edge of science and performance. Today's greatest innovations are tomorrow's baseline, and you have to keep moving forward. That's the only way to continue our physical and intellectual growth as a species. So are we humans like Thoroughbreds or greyhounds, approaching the edge of our abilities?

Maybe that's not even the right question to ask. Let me leave you with this thought: It's not just that we're trying to push the frontiers of our current understanding, but that we're trying to discover new frontiers altogether. When I was sitting with the Red Bull sports science team, watching them pore over all the data they had gathered on their athletes, looking for a tiny edge that might help one of them win a competition, I asked about the limits of what we can do as a species. Surely we'll never run a marathon in twenty minutes, I said, or clock a hundred meters in eight seconds.

Walshe looked up at me.

"What's the potential for human beings if you focus on the far future?" he asked. "If you look back thirty thousand or forty thousand years, how were Neanderthals running around then? And what if you reverse that logic? The idea is that there are potentially multiple thousands of a percent of improvement available to us. And that we're not talking about that happening in thousands of years, but hundreds. There's unlimited potential here, we think, long term.

"So you ask, Will someone ever run a hundred meters in eight seconds? Yes. Will they run it in six? Yes, but those athletes might not look like anything we can even imagine today."

There's a wonder and optimism there—a sense that we're still just scratching the surface of what we can do, not just with our bodies but also with our minds. Walshe smiles at me as I digest what he's just said, turning over the implications. I try to picture the far future that he describes, and think of all the scientists and athletes I've met who are working so hard to make it come to fruition.

I just can't wait to see it.

ACKNOWLEDGMENTS

First, a sincere thank-you to all the athletes, coaches, scientists, historians, and researchers who took the time to talk to me during my reporting on the intersection of sports and science, both those who are quoted directly in the book as well as those who helped shape my thinking and ideas through our conversations. A journalist is only as good as the time and generosity of his sources allow him to be, and these people gave greatly to help me understand their world:

Gary Abbott, Hank Adams, Gil Blander, Gretchen Bleiler, Chris Boardman, Phil Borgeaud, Claude Bouchard, Liz Broad, George Brooks, Phil Brown, Mark Denny, Scott Drawer, Ashton Eaton, John Fowlie, Clint Friesen, Kirk Goldsberry, Christopher Gore, Jason Gulbin, Shona Halson, Tobie Hatfield, Craig Hilliard, Will Hopkins, Hamish Jeacoke, Andrew Jones, Lolo Jones, Asker Jeukendrup, John Kessel, Stuart Kim, Richard Kirby, Floyd Landis, Ted Ligety, Carl Lewis, Per Lundstam, Holden MacRae, Samuele Marcora, David Martin, Bruce Mason, Bill Moreau, Tim Noakes, Cuan Petersen, Dick Pound, James Roberts, Matt

Schiller, Geoff Shackelford, Dennis Shaver, Adir Shiffman, Philip Skiba, Ross Smith, Rohan Taylor, Ross Tucker, Benoit Vincent, Peter Vint, Phil Wagner, Andy Walshe, Randy Wilber, and Mounir Zok.

Special thanks to David Packwood at the Australian Institute of Sport, Brittany Davis of the U.S. Olympic Committee, Katie Denbo at Taylor-Made Golf, Ilana Taub and Erin Mand at Red Bull, Joe Gomes at Titleist, U.S. Ski Team's Doug Haney, and Matt Daniels at Doyle Management for their help in arranging visits to their companies or organizations and helping coordinate interviews. Also, thank you to the library staff at the University of California, Berkeley—especially at the Bioscience and Natural Resources Library—as well as the staff at the University of California, San Francisco, library for their help with my research.

I've found a professional home at *WIRED*, where my colleagues make coming to work a true pleasure every day. Some specific thanks: first to Chris Anderson and Thomas Goetz for bringing me into the fold, and for their early support of this project. Evan Hansen offered me a chance to write about sports and technology at WIRED.com, which I'm now proud to edit. Scott Dadich, Jason Tanz, and Jacob Young helped me juggle two jobs through the completion of this book, and Adam Rogers and Robert Capps were crucial not only as sounding boards for ideas but also for their friendship and encouragement. Robert also introduced me to Phil Wagner at Sparta Performance Science, for which I'm grateful.

Portions of this book have previously appeared in *WIRED* and on WIRED.com as parts of coverage I've done of sports, science, and technology over the past nine years. My thanks to *WIRED* and Condé Nast to allow its use here—I've reworked some passages, while some remain largely as they were written at the time.

Every writer should be as lucky as I am to have an agent as smart and thoughtful as David Fugate. David sought me out after reading one of my stories in *WIRED*, and spent countless hours helping me shape my ideas and thinking. At Hudson Street Press, I've been privileged to work with Caroline Sutton and Christina Rodriguez, who turned this into a much better-argued and -written book than it would have been without their formidable talents.

One day, my daughter Paige turned to me and said, "Daddy, what if you make a mistake in the book and then the whole thing is ruined?" I told her that I would try very hard not to make any mistakes at all—and I have to thank Katie Palmer for all the research and fact-checking help in my own pursuit of perfection.

My daughters, Kate and Paige, have been fascinated with the idea of this book. "Is it going to be a real book?" they asked. I hope so, I said. "Will your name be on the cover?" they asked me. I assured them it would. They're two of the most voracious readers I know, and I really look forward to them getting a chance to dive into this world that's consumed their dad for so long. Love you, Kate-o and Wiggle Pop.

And finally, the largest thanks goes to my wife, Kristen. Throughout the long and often difficult process of writing this book, the duties that I shirked at home landed squarely on her. She took on the additional burdens with a smile, and offered me constant encouragement throughout. I hope the results repay, in some small way, what she's done for me. Her support, her perspective, her humor, her love—they were a daily reminder of how lucky I am to have found her, and how having her in my life doesn't just make it better. It makes it possible.

NOTES

Chapter 1: Hacking the Athlete

1 **decrease an athlete's strength and explosiveness:** K. Power, D. Behm, F. Cahill, M. Carroll, and W. Young, "An Acute Bout of Static Stretching: Effects on Force and Jumping Performance," *Medicine & Science in Sports & Exercise* 36, no. 8 (2004): 1389–96.

6 **"Performance by the aggregation of marginal gains":** Richard Moore, *Heroes, Villains & Velodromes* (London: HarperSport, 2012), 247.

7 **"more and more by scientific and technological advances":** Giuseppe Lippi, Giuseppe Banfi, Emmanuel J. Favaloro, Joern Rittweger, and Nicola Maffulli, "Updates on Improvement of Human Athletic Performance: Focus on World Records in Athletics," *British Medical Bulletin* 87 (2008): 14.

7 **suffered from some sort of illness:** Lars Engebretsen, Torbjørn Soligard, Kathrin Steffen, Juan Manuel Alonso, Mark Aubry, Richard Budgett, Jiri Dvorak, et al., "Sports Injuries and Illnesses During the London Summer Olympic Games 2012," *British Journal of Sports Medicine* 47, no. 7 (2013): 407–14.

7 **"And if you look at all the teams' data":** Alasdair Frotheringham, "'Head of Marginal Gains' Helps GB Gold Machine Stay in Front," *Independent*, August 7, 2012, http://www.independent.co.uk/sport/olympics/cycling/head-of-marginal-gains-helps-gb-gold-machine-stay-in-front-8010110.html.

9 **from Usain Bolt on down to you and me:** P. G. Weyand, D. B. Sternlight, M. J. Bellizzi, and S. Wright, "Faster Top Running Speeds Are Achieved with Greater Ground Forces Not More Rapid Leg Movements," *Journal of Applied Physiology* 89, no. 5 (2000): 1991–9.

9   **"Speed is conferred predominantly":** Ibid.
9   **other endurance athletes after strength training:** K. Beattie, I. C. Kenny, M. Lyons, and B. P. Carson, "The Effect of Strength Training on Performance in Endurance Athletes," *Sports Medicine*, published online before print (2014), http://dx.doi.org/10.1007/s40279-014-0157-y.
9   **squats in the weight room at six hundred pounds or more:** Ralph Mann, "The Phases of Sprinting," presentation at the USATF/VS Athletics Trials Super Clinic, Eugene, Oregon (June 26, 2012).

Chapter 2: Gold Medal Genetics

16   **the quality of their nutrition:** R. Makowsky, N. M. Pajewski, Y. C. Klimentidis, A. I. Vazquez, C. W. Duarte, D. B. Allison, and G. de los Campos, "Beyond Missing Heritability: Prediction of Complex Traits," *PLoS Genetics* 7, no. 4 (2011): e1002051, doi:10.1371/journal.pgen.1002051.
17   **"The overall objective of the HERITAGE project":** Claude Bouchard, A. S. Leon, D. C. Rao, J. S. Skinner, J. H. Wilmore, and J. Gagnon, "The HERITAGE Family Study. Aims, Design, and Measurement Protocol," *Medicine & Science in Sports & Exercise* 27, no. 5 (1995): 722.
18   **"The probability of this pattern of familial selection":** Kevin Norton and Tim Olds, "Morphological Evolution of Athletes Over the 20th Century: Causes and Consequences," *Sports Medicine* 31, no. 11 (2001): 780.
18   **"genetic polymorphisms of the next generation of gifted athletes":** Ibid.
18   **Men and women differ in several physical and physiological ways:** W. Larry Kenney, Jack H. Wilmore, and David L. Costill, *Physiology of Sport and Exercise*, 5th ed. (Champaign, IL: Human Kinetics, 2011).
19   **French researchers built a database of the world records:** Valérie Thibault, M. Guillaume, G. Berthelot, N. E. Helou, K. Schaal, L. Quinquis, H. Nassif, et al., "Women and Men in Sport Performance: The Gender Gap Has Not Evolved Since 1983," *Journal of Sports Science and Medicine* 9, no. 2 (2010): 214–23.
19   **"Without any technological improvement specifically dedicated to one gender":** Ibid., 221.
20   **Kenyan and Ethiopian runners account for 90 percent:** Randall L. Wilber and Yannis P. Pitsiladis, "Kenyan and Ethiopian Distance Runners: What Makes Them So Good?" *International Journal of Sports Physiology and Performance* 7, no. 2 (2012): 92–102.
22   **associated with increased performance in sprinting or power sports:** Elaine A. Ostrander, Heather J. Huson, and Gary K. Ostrander, "Genetics of Athletic Performance," *Annual Review of Genomics and Human Genetics* 10 (2009): 407–29.
22   **Those observed benefits of the ACE insertion have all been in Caucasian athletes:** Ross Tucker, Jordan Santos-Concejero, and Malcolm Collins, "The Genetic

Basis for Elite Running Performance," *British Journal of Sports Medicine* 47, no. 9 (2013): 545–9.

22 **The results of the first study on this, in 2003:** N. Yang, D. G. MacArthur, J. P. Gulbin, A. G. Hahn, A. H. Beggs, S. Easteal, and K. North, "ACTN3 Genotype Is Associated with Human Elite Athletic Performance," *American Journal of Human Genetics* 73, no. 3 (2003): 627–31.

23 **the country only contained 2 percent of people with the XX variant:** R. A. Scott, R. Irving, L. Irwin, E. Morrison, V. Charlton, K. Austin, D. Tladi, et al., "ACTN3 and ACE Genotypes in Elite Jamaican and US Sprinters," *Medicine & Science in Sports & Exercise* 42, no. 1 (2010): 107–12.

25 **VO_2 max isn't a particularly great predictor of performance:** M. J. Joyner, J. R. Ruiz, and A. Lucia, "The Two-Hour Marathon: Who and When?" *Journal of Applied Physiology* 110, no. 1 (2011): 275–7.

26 **her VO_2 max ranged between 65 and 75:** Andrew M. Jones, "The Physiology of the World Record Holder for the Women's Marathon," *International Journal of Sports Science and Coaching* 1, no. 2 (2006): 101–16.

26 **"a prerequisite for successful performance at the international level":** Ibid., 108.

26 **Half of our native ability to process oxygen:** Yannis P. Pitsiladis (Chair), K. Anders Ericsson, Phillip Ackerman, and Claude Bouchard, "The Genetics Talent Myth and the 10,000 Hour Rule," presentation at the annual meeting of the American College of Sports Medicine, San Francisco (June 1, 2012).

28 **a 15 percent increase in her running economy:** Jones, "Physiology of the World Record Holder for the Women's Marathon."

29 **In fact, even the best coaches:** Gina Kolata, "Running Efficiency: It's Good, But How Do You Get It?" *The New York Times*, October 11, 2007, http://www.nytimes .com/2007/10/11/fashion/11Best.html.

30 **all those SNPs combined accounted for nearly half of the observed variance:** J. Yang, B. Benyamin, B. P. McEvoy, S. Gordon, A. K. Henders, D. R. Nyholt, P. A. Madden, et al., "Common SNPs Explain a Large Proportion of the Heritability for Human Height," *Nature Genetics* 42, no. 7 (2010): 565–9.

30 **"Athletic performance is undoubtedly more complex than height":** Ross Tucker and Malcolm Collins, "What Makes Champions? A Review of the Relative Contribution of Genes and Training to Sporting Success," *British Journal of Sports Medicine* 46, no. 8 (2012): 557.

31 **gaining almost three times as much aerobic fitness as the low-responding group:** C. Bouchard, M. A. Sarzynski, T. K. Rice, W. E. Kraus, T. S. Church, Y. J. Sung, D. C. Rao, and T. Rankinen, "Genomic Predictors of the Maximal O_2 Uptake Response to Standardized Exercise Training Programs," *Journal of Applied Physiology* 110, no. 5 (2011): 1160–70.

31 **Bouchard's lab found that around 10 percent of participants:** Claude Bouchard, S. N. Blair, T. S. Church, C. P. Earnest, J. M. Hagberg, K. Häkkinen, N.

T. Jenkins, et al., "Adverse Metabolic Response to Regular Exercise: Is It a Rare or Common Occurrence?" *PloS One* 7, no. 5 (2012): e37887, doi:10.1371/journal .pone.0037887.

32 **deficiencies in vitamins and minerals:** Gabrielle T. Goodlin, Andrew K. Roos, Thomas R. Roos, Claire Hawkins, Sydney Beache, Stephen Baur, and Stuart K. Kim, "Applying Personal Genetic Data to Injury Rick Assessment in Athletes" (working paper, Stanford University, 2014).

33 **effects from doping that persist even after an athlete stops taking the drugs:** Tobias Ehlert, Perikles Simon, and Dirk A. Moser, "Epigenetics in Sports," *Sports Medicine* 43, no. 2 (2013): 93–110.

Chapter 3: Eating to Win

38 **The typical athlete, without using any special nutritional techniques:** Louise M. Burke and John A. Hawley, "Effects of Short-Term Fat Adaptation on Metabolism and Performance of Prolonged Exercise," *Medicine and Science in Sports and Exercise* 34, no. 9 (2002): 1492–8.

39 **In fact, Asker Jeukendrup, the sports scientist:** Asker E. Jeukendrup, "Nutrition for Endurance Sports: Marathon, Triathlon, and Road Cycling," *Journal of Sports Sciences* 29, suppl. 1 (2011): S91–9.

40 **up to three liters an hour in one study of a 2:06 marathoner:** L. Y. Beis, M. Wright-Whyte, B. Fudge, T. Noakes, and Y. P. Pitsiladis, "Drinking Behaviors of Elite Male Runners During Marathon Competition," *Clinical Journal of Sport Medicine* 22, no. 3 (2012): 254–61.

41 **top sports nutritionists suggest that you don't need any carbs whatsoever:** L. M. Burke, J. A. Hawley, S. H. Wong, and A. E. Jeukendrup, "Carbohydrates for Training and Competition," *Sports Medicine* 29, suppl. 1 (2011): S17–27.

41 **the recommendation is to consume between 30 and 60 grams of carbs per hour:** Ibid.

42 **the same nutritional profile as the versions in the grocery store:** "Nancy Clark's Homemade Sports Drink," U.S. Olympic Committee (2010), http://www.team usa.org/~/media/TeamUSA/sport%20performance/pdf%20handouts/Nancy _Clark_s_Sports_Drink.pdf.

42 **the sort of intake that Tour de France riders strive for:** Burke et al., "Carbohydrates for Training and Competition."

43 **The AIS has set up a website:** Australian Institute of Sport, "Supplements," accessed February 23, 2014, http://www.ausport.gov.au/ais/nutrition/supplements.

44 **cause inflammation and reduce the effectiveness of muscles:** Enette Larson-Meyer, "Vitamin D Supplementation in Athletes," *Nestlé Nutrition Institute Workshop Series* 75 (2013): 109–21.

45   **removed from the banned list in 2004:** S. J. Stear, L. M. Castell, L. M. Burke, and L. L. Spriet, "BJSM Reviews: A–Z of Nutritional Supplements: Dietary Supplements, Sports Nutrition Foods and Ergogenic Aids for Health and Performance Part 6," *British Journal of Sports Medicine* 44, no. 4 (2010): 297–8.

45   **"Caffeine supplementation is likely to be beneficial":** Ibid., 297.

46   **a very minor effect on their hydration status:** Lawrence E. Armstrong, "Caffeine, Body Fluid-Electrolyte Balance, and Exercise Performance," *International Journal of Sport Nutrition and Exercise Metabolism* 12, no. 2 (2002): 189–206.

46   **That's because hydrogen ions are released:** Lars R. McNaughton, Jason Siegler, and Adrian Midgley, "Ergogenic Effects of Sodium Bicarbonate," *Current Sports Medicine Reports* 7, no. 4 (2008): 230–6.

47   **0.3 grams of sodium bicarbonate per kilogram of body weight:** Amelia J. Carr, Will G. Hopkins, and Christopher J. Gore, "Effects of Acute Alkalosis and Acidosis on Performance: A Meta-analysis," *Sports Medicine* 41, no. 10 (2011): 801–14.

47   **kept many athletes who could benefit from sodium bicarbonate from using it at all:** A. J. Carr, G. J. Slater, C. J. Gore, B. Dawson, and L. M. Burke, "Effect of Sodium Bicarbonate on [HCO3-], pH, and Gastrointestinal Symptoms," *International Journal of Sport Nutrition and Exercise Metabolism* 21, no. 3 (2011): 189–94.

48   **lifters taking creatine improved as much as 14 percent more:** R. Cooper, F. Naclerio, J. Allgrove, and A. Jimenez, "Creatine Supplementation with Specific View to Exercise/Sports Performance: An Update," *Journal of the International Society of Sports Nutrition* 9, no. 1 (2012): 33.

49   **important to endurance exercise:** Ibid.

49   **reducing oxygen consumption in submaximal efforts:** Andrew M. Jones, Helen Carter, Jamie S. M. Pringle, and Iain T. Campbell, "Effect of Creatine Supplementation on Oxygen Uptake Kinetics During Submaximal Cycle Exercise," *Journal of Applied Physiology* 92, no. 6 (2002): 2571–7.

49   **no one has been able to establish any health risks:** J. R. Poortmans, E. S. Rawson, L. M. Burke, S. J. Stear, and L. M. Castell, "A–Z of Nutritional Supplements: Dietary Supplements, Sports Nutrition Foods and Ergogenic Aids for Health and Performance Part 11," *British Journal of Sports Medicine* 44, no. 10 (2010): 765–6.

50   **in events that last between five and thirty minutes:** K. E. Lansley, P. G. Winyard, S. J. Bailey, A. Vanhatalo, D. P. Wilkerson, J. R. Blackwell, M. Gilchrist, N. Benjamin, and A. M. Jones, "Acute Dietary Nitrate Supplementation Improves Cycling Time Trial Performance," *Medicine & Science in Sports & Exercise* 43, no. 6 (2011): 1125–31.

50   **The results are more ambiguous for longer events:** D. P. Wilkerson, G. M.

Hayward, S. J. Bailey, A. Vanhatalo, J. R. Blackwell, and A. M. Jones, "Influence of Acute Dietary Nitrate Supplementation on 50 Mile Time Trial Performance in Well-Trained Cyclists," *European Journal of Applied Physiology* 112, no. 12 (2012): 4127–34.

52 **raise the level of carnosine in muscles by 40 to 60 percent:** R. C. Harris, M. J. Tallon, M. Dunnett, L. Boobis, J. Coakley, H. J. Kim, J. L. Fallowfield, C. A. Hill, C. Sale, and J. A. Wise, "The Absorption of Orally Supplied Beta-Alanine and Its Effect on Muscle Carnosine Synthesis in Human Vastus Lateralis," *Amino Acids* 30, no. 3 (2006): 279–89.

52 **In a 2012 meta-analysis of research on beta-alanine:** R. M. Hobson, B. Saunders, G. Ball, R. C. Harris, and C. Sale, "Effects of ß-Alanine Supplementation on Exercise Performance: A Meta-analysis," *Amino Acids* 43, no. 1 (2012): 25–37.

52 **"One could term the pH buffers inside the muscle cells":** Asker E. Jeukendrup, "Performance and Endurance in Sport: Can It All Be Explained by Metabolism and Its Manipulation?" *Dialogues in Cardiovascular Medicine* 17 (2012): 43.

52 **more effective together than separately:** Craig Sale, Bryan Saunders, Sean Hudson, John A. Wise, Roger C. Harris, and Caroline D. Sunderland, "Effect of Beta-Alanine Plus Sodium Bicarbonate Supplementation on High-Intensity Cycling Capacity," *Medicine & Science in Sports & Exercise* 42, no. 10 (2011): 1972–8.

54 **That leads to a greater response to the same training stimulus:** Louise M. Burke, "New Issues in Training and Nutrition: Train Low, Compete High?" *Current Sports Medicine Reports* 6, no. 3 (2007): 137–8.

54 **It was an ingeniously designed experiment:** A. K. Hansen, C. P. Fischer, P. Plomgaard, J. L. Andersen, B. Saltin, and B. K. Pedersen, "Skeletal Muscle Adaptation: Training Twice Every Second Day vs. Training Once Daily," *Journal of Applied Physiology* 98, no. 1 (2005): 93–9.

54 **not a clear line between those molecular adaptations:** Louise M. Burke, "Fueling Strategies to Optimize Performance: Training High or Training Low?" *Scandinavian Journal of Medicine & Science in Sports* 20, suppl. 2 (2010): 48–58.

55 **"Reasons for this apparent disconnect":** Ibid., 56.

59 **"we have not effectively used the abilities science":** Atul Gawande, *Better: A Surgeon's Notes on Performance* (New York: Metropolitan Books, 2007), 232.

Chapter 4: The Search for Excellence

62 **David Nadel in a newspaper interview:** Samantha Lane, "Preoccupied with Obscure Olympic Games," *Sydney Morning Herald*, April 20, 2012, http://www .smh.com.au/sport/olympics-2012/preoccupied-with-obscure-olympic-games -20120419-1xaaq.html.

65 **weren't among the best in the world when they were younger:** Joshua L. Foss

and Robert F. Chapman, "Career Performance Progressions of Junior and Senior Elite Track and Field Athletes," poster session at the annual meeting of the American College of Sports Medicine, Indianapolis (May 28–June 1, 2013).

65 **"There is no association between rankings":** Juliane Wulff and Antje Hoffmann, "Competition Results—Appropriate Criteria for Talent Selection in Triathlon?" poster session at the annual meeting of the European College of Sport Science, Barcelona, Spain (June 26–29, 2013).

66 **"Elite performers in senior sports":** Indiana University, "Elite Athletes Often Shine Sooner or Later—But Not Both," *ScienceDaily*, May 31, 2013, www.science daily.com/releases/2013/05/130531105413.htm.

66 **kids born in the first half of the year have more success:** Jochen Musch and Simon Grondin, "Unequal Competition as an Impediment to Personal Development: A Review of the Relative Age Effect in Sport," *Developmental Review* 21, no. 2 (2001): 147–67.

67 **roughly four times as many top junior hockey players:** Roger H. Barnsley and A. H. Thompson, "Birthdate and Success in Minor Hockey: The Key to the NHL," *Canadian Journal of Behavioural Science* 20, no. 2 (1988): 167–76.

67 **"relatively younger athletes are more likely to be chosen":** Joseph Baker and A. Jane Logan, "Developmental Contexts and Sporting Success: Birth Date and Birthplace Effects in National Hockey League Draftees 2000–2005," *British Journal of Sports Medicine* 41, no. 8 (2007): 516.

67 **a study of the salaries of players in Germany's top soccer league:** John Ashworth and Bruno Heyndels, "Selection Bias and Peer Effects in Team Sports: The Effect of Age Grouping on Earnings of German Soccer Players," *Journal of Sports Economics* 8, no. 4 (2007): 355–77.

68 **"The U.S. data indicate that children born":** J. Côté, D. J. Macdonald, J. Baker, and B. Abernethy, "When 'Where' Is More Important Than 'When': Birthplace and Birthdate Effects on the Achievement of Sporting Expertise," *Journal of Sports Sciences* 24, no. 10 (2006): 1070.

69 **In 2012, 37 percent of black children:** U.S. Department of Health and Human Services, "Information on Poverty and Income Statistics: A Summary of 2013 Current Population Survey Data," September 2013, http://aspe.hhs.gov/hsp/13/ PovertyAndIncomeEst/ib_poverty2013.cfm.

70 **"Looking at the combined categories of upper and middle class of origin":** Joshua Kjerulf Dubrow and Jimi Adams, "Hoop Inequalities: Race, Class and Family Structure Background and the Odds of Playing in the National Basketball Association," *International Review for the Sociology of Sport* 47, no. 1 (2012): 43–59.

70 **There was one substantial difference in Dubrow and Adams's study:** Ibid.

70 **anger and resentment is essential fuel for many athletes:** Kevin G. Thompson (Chair), Alan St. Clair Gibson, and Jack Raglin, "What Creates a Sporting Genius—

New Concepts in Talent ID," presentation at the annual meeting of the American College of Sports Medicine, San Francisco (May 31, 2012).

71 **"You have to be a masochist":** Jim White, "Mark Cavendish admits: 'You have to be a masochist to succeed in Tour de France'," *The Telegraph*, April 19, 2010, http://www.telegraph.co.uk/sport/othersports/cycling/mark-cavendish/7607784/Mark-Cavendish-admits-You-have-to-be-a-masochist-to-succeed-in-Tour-de-France.html.

71 **"DNA donor":** Lance Armstrong with Sally Jenkins. *It's Not About the Bike: My Journey Back to Life* (New York: Putnam, 2000), 20.

71 **"The knowledge and skills the athletes accrued from 'life' traumas":** Dave Collins and Áine MacNamara, "The Rocky Road to the Top: Why Talent Needs Trauma," *Sports Medicine* 42, no. 11 (2012): 909.

73 **"They are known to take responsibility for the progress they make":** M. T. Elferink-Gemser, G. Jordet, M. J. Coelho-E-Silva, and C. Visscher, "The Marvels of Elite Sports: How to Get There?" *British Journal of Sports Medicine* 45, no. 9 (2011): 683–4.

75 **Duffield has found that at 100 meters:** Rob Duffield, Brian Dawson, and Carmel Goodman, "Energy System Contribution to 100-m and 200-m Track Running Events," *Journal of Science and Medicine in Sport* 7, no. 3 (2004): 302–13.

75 **For the 3,000 meters, that's roughly flipped:** Rob Duffield, Brian Dawson, and Carmel Goodman, "Energy System Contribution to 1500- and 3000-metre Track Running," *Journal of Sports Sciences* 23, no. 10 (2005): 993–1002.

Chapter 5: The Fast Track to Greatness

79 **"In the history of the Olympics, forty-two athletes":** Timothy Olds, "Body Composition and Sports Performance," in *The Olympic Textbook of Science in Sport*, ed. Ronald J. Maughan (Oxford: Wiley-Blackwell, 2009), 137.

80 **Height is most advantageous during the serve:** Ibid., 139.

80 **In 2011, *Sports Illustrated* estimated:** Pablo S. Torre, "Larger Than Real Life," *Sports Illustrated*, July 4, 2011, http://sportsillustrated.cnn.com/vault/article/magazine/MAG1187806/index.htm.

80 **rowing is one of the most morphologically constrained sports:** Olds, "Body Composition and Sports Performance," 135.

81 **"I remember sitting in a room in Bisham Abbey":** James Skitt, "Team GB Olympic Gold Medallist Unearthed by UK Sport's Talent Scheme," *UK Sport*, August 1, 2012, http://www.uksport.gov.uk/news/team-gb-olympic-gold-medallist-unearthed-by-uk-sport-talent-scheme/.

81 **In the first year, seventeen of the athletes identified:** Roel Vaeyens, A. Güllich, C. R. Warr, and R. Philippaerts, "Talent Identification and Promotion Pro-

grammes of Olympic Athletes," *Journal of Sports Sciences* 27, no. 13 (2009): 1367–80.

82 **"Many characteristics once believed to reflect innate talent":** K. Anders Ericsson, Ralf Th. Krampe, and Clemens Tesch-Römer, "The Role of Deliberate Practice in the Acquisition of Expert Performance," *Psychological Review* 100, no. 3 (1993): 363.

82 **"The central claim of our framework":** Ibid., 370.

82 **"The idea that excellence at performing a complex task":** Malcolm Gladwell, *Outliers: The Story of Success* (New York: Little, Brown and Company, 2008), 39.

83 **"That was Thursday, Jan. 19, 2006":** Brett Hess, "Just Like the Movies," March 20, 2007, http://www.cstv.com/sports/c-track/stories/032007aax.html.

84 **For the wrestlers, it was about 6,000 hours:** Nicola J. Hodges and Janet L. Starkes, "Wrestling with the Nature Expertise: A Sport Specific Test of Ericsson, Krampe and Tesch-Römer's (1993) Theory of 'Deliberate Practice,'" *International Journal of Sport Psychology* 27, no. 4 (1996): 400–24.

84 **the field hockey players took about 4,000 hours:** Werner F. Helsen, Janet L. Starkes, and Nicola J. Hodges, "Team Sports and the Theory of Deliberate Practice," *Journal of Sport & Exercise Psychology* 20 (1998), 12–34.

84 **A 2004 study looked at the developmental history:** Karen Oldenziel, Françoys Gagné, and Jason P. Gulbin, "Factors Affecting the Rate of Athlete Development from Novice to Senior Elite: How Applicable Is the 10-Year Rule?" paper presented to the Pre-Olympic Congress *Sports Science Through the Ages*, Athens, Greece (August 2004).

85 **"Consistent with our hypothesis":** Ericsson, Krampe, and Tesch-Römer, "The Role of Deliberate Practice," 389.

86 **Athletes who exceeded this guideline were 70 percent:** Loyola Health University System, "Intense, Specialized Training in Young Athletes Linked to Serious Overuse Injuries," ScienceDaily, April 19, 2013, www.sciencedaily.com/releases/2013/04/130419132508.htm.

87 **"highest when the athletes had experienced more sports":** A. Güllich and E. Emrich, "Considering Long-term Sustainability in the Development of World Class Success," *European Journal of Sport Science*, 14, suppl. 1, (2014): S383-97.

87 **"deliberate practice is *sufficient* to explain the acquisition":** K. Anders Ericsson, "Training History, Deliberate Practice and Elite Sports Performance: An Analysis in Response to Tucker and Collins Review—What Makes Champions?" *British Journal of Sports Medicine* 47, no. 9 (2013): 533–5.

87 **"the best group of violinists had spent significantly more hours":** Ibid., 534.

88 **"Skeleton offered a unique opportunity to identify a complete novice":** N. Bullock, J. P. Gulbin, D. T. Martin, A. Ross, T. Holland, and F. Marino, "Talent Identification and Deliberate Programming in Skeleton: Ice Novice to Winter Olympian in 14 Months," *Journal of Sports Sciences* 27, no. 4 (2009): 398.

89 **In the paper describing the skeleton program:** Ibid., 403.

90 **"They said I had one of their best results ever":** João Medeiros, "Winning by Numbers: How Performance Analysis Is Transforming Sport," *WIRED UK*, July 2012.

91 **"People talk about the ten years it takes":** Ibid., 127.

91 **"The activation of genes is critical for developing physiological adaptations":** K. Anders Ericsson, Kiruthiga Nandagopal, and Roy W. Roring, "Toward a Science of Exceptional Achievement: Attaining Superior Performance Through Deliberate Practice," *Annals of the New York Academy of Sciences* 1172 (2009): 212.

94 **"Genes determine the size of the bucket":** Richard C. Lewontin, *The Triple Helix: Gene, Organism, and Environment* (Cambridge, MA: Harvard University Press, 2000), 26.

94 **"training maximises the likelihood of obtaining a performance level":** Tucker and Collins, "What Makes Champions?" 560.

Chapter 6: Learning to Be the Best

101 **demonstrated by a clever experiment published in 1979:** John B. Shea and Robyn L. Morgan, "Contextual Interference Effects on the Acquisition, Retention, and Transfer of a Motor Skill," *Journal of Experimental Psychology: Human Learning and Memory* 5, no. 2 (1979): 179–87.

104 **"We were not surprised to see athletes' initial scores":** Jocelyn Faubert, "Professional Athletes Have Extraordinary Skills for Rapidly Learning Complex and Neutral Dynamic Visual Scenes," *Scientific Reports* 3 (2013): 1154, doi:10.1038/srep01154.

106 **"Strength training is a great short-term investment":** G. Martin Bingisser, "Simplifying Bondarchuk," *Modern Athlete & Coach*, April 2010.

108 **"Futsal compresses soccer's essential skills into a small box":** Daniel Coyle, *The Talent Code: Greatness Isn't Born. It's Grown. Here's How* (New York: Bantam Dell, 2009), 27.

108 **"The small playing area helped me improve my close control":** "The Football Greats Forged by Futsal," FIFA.com, October 30, 2012, http://www.fifa.com/futsalworldcup/news/newsid=1798909/index.html.

109 **"It was huge in developing my passing touch":** Kelly Whiteside, "Draft's Top QBs Honed Skills in Texas 7-on-7," *USA Today*, April 25, 2012, http://usatoday30.usatoday.com/sports/football/nfl/story/2012-04-25/luck-griffin-tannehill-texas-7-on-7/54542400/1.

110 **"The way we learn in residency currently":** American College of Surgeons, "Novel 3-D Simulation Technology Helps Surgical Residents Train More Effectively," ScienceDaily, August 2, 2013, http://www.sciencedaily.com/releases/2013/08/130802095148.htm.

112 **"These games nowadays are just so technically sound":** Chris Suellentrop, "Game Changers: How Videogames Trained a Generation of Athletes," *WIRED*, February 2010.

Chapter 7: Tools of the Trade

117 **"Today I finally had the chance to try a prototype":** Ted Ligety, "My Thoughts on FIS's Attempt to Ruin GS," Ted Ligety, Alpine Ski Racer, August 18, 2011, http://www.tedligety.com/blog/my-thoughts-on-fiss-attempt-to-ruin-gs/.

120 **"the spirit of fair play":** IOC Executive Board, *Decision Regarding Mr. Ara Abrahamian* (2008), http://www.olympic.org/Documents/Reports/EN/en_report_1354 .pdf.

125 **"We didn't have a clue":** Jonathan Wall, "The Evolution of Titleist's Pro V1," PGATour.com, September 6, 2013, http://www.pgatour.com/news/2013/09/06/ the-evolution-of-the-pro-v1.html.

129 **there's a page at the UK Sport website:** UK Sport, "Cycling," accessed November 24, 2013, http://www.uksport.gov.uk/pages/cycling/.

130 **A second frame, the MK2:** Dimitris Katsanis, "The Olympic Carbon Bicycle Frame," presentation at the Subcon 2013, Birmingham, England (2013).

132 **"What is surprising to people is that":** Tim Newcomb, "U.S. Speedskating Finds Edge with High-Tech Engineered Skins," SI.com, January 16, 2014, http:// sportsillustrated.cnn.com/-olympics/news/20140116/sochi-olympics-speed -skating-under-armour-mach-39-skin/.

133 **The *Wall Street Journal* brought the story to light:** Joshua Robinson and Sara Germano, "Sochi Olympics: Under Armour Suits May Be a Factor in U.S. Speedskating's Struggles," WSJ.com, February 13, 2014, http://online.wsj.com/ news/articles/SB10001424052702304703804579381002780722432.

134 **As a 2013 editorial on the role of the placebo effect:** Shona L. Halson and David T. Martin, "Lying to Win—Placebos and Sport Science," *International Journal of Sports Physiology and Performance* 8, no. 6 (2013): 597–9.

Chapter 8: What Getting Tired Means

139 **He won a Nobel Prize in Physiology or Medicine in 1922:** David R. Bassett Jr., "Scientific Contributions of A. V. Hill: Exercise Physiology Pioneer," *Journal of Applied Physiology* 93, no. 5 (2002): 1567–82.

140 **which would project out to a sub-three-hour marathon:** Ibid.

140 **lactic acid would build up, causing fatigue:** A. V. Hill and Hartley Lupton, "Muscular Exercise, Lactic Acid, and the Supply and Utilization of Oxygen," *QJM* os-16, no. 62 (1923): 135–71.

141 **"Were it not for the fact that the body":** Ibid., 137.

141 **an inability to do any more work:** T. D. Noakes and A. St. Clair Gibson, "Logical Limitations to the 'Catastrophe' Models of Fatigue During Exercise in Humans," *British Journal of Sports Medicine* 38, no. 5 (2004): 648–9.

141 **The first edition of Dr. George Brooks's textbook:** George Austin Brooks and Thomas Davin Fahey, *Exercise Physiology: Human Bioenergetics and Its Applications* (New York: John Wiley & Sons, 1984).

142 **Meyerhof found a high amount of a compound:** Robert A. Roberts, Farzenah Ghiasvand, and Daryl Parker, "Biochemistry of Exercise-Induced Metabolic Acidosis," *American Journal of Physiology: Regulatory, Integrative and Comparative Physiology* 287, no. 3 (2004): R502–16.

145 **Brooks found that the trained riders could produce:** L. A. Messonnier, C. A. Emhoff, J. A. Fattor, M. A. Horning, T. J. Carlson, and G. A. Brooks, "Lactate Kinetics at the Lactate Threshold in Trained and Untrained Men," *Journal of Applied Physiology* 114, no. 11 (2013): 1593–602.

146 **"experiencing heaven on the road":** Tim Noakes and Michael Vlismas, *Challenging Beliefs: Memoirs of a Career* (Cape Town, South Africa: Zebra Press, 2011), 25.

147 **"Hill's original conclusions were not supported by his own findings":** Timothy David Noakes, "Challenging Beliefs: Ex Africa Semper Aliquid Novi," *Medicine & Science in Sports & Exercise* 29, no. 5 (1997): 571–90.

149 **That's the end spurt at the very top level of the athletic world:** Ross Tucker, Michael I. Lambert, and Timothy D. Noakes, "An Analysis of Pacing Strategies During Men's World-Record Performances in Track Athletics," *International Journal of Sports Physiology and Performance* 1, no. 3 (2006): 233–45.

149 **Some research argues that athletes can refill:** Philip F. Skiba, Sarah Jackman, David Clarke, Anni Vanhatalo, Andrew M. Jones, "Effect of Work and Recovery Durations on Reconstitution During Intermittent Exercises," *Medicine & Science in Sports & Exercise* (January 30, 2014), doi:10.1249/mss.0000000000000226.

149 **that nothing in our body is pushed beyond the normal range:** Noakes and Vlismas, *Challenging Beliefs*, 254.

150 **"We propose that fatigue is a combination of the brain":** Ibid., 266.

151 **When the cyclists rinsed their mouths with the carb solution:** James M. Carter, Asker E. Jeukendrup, and David A. Jones, "The Effect of Carbohydrate Mouth Rinse on 1-h Cycle Time Trial Performance," *Medicine & Science in Sports & Exercise* 36, no. 12 (2004): 2107–11.

152 **"The fact that the subjects in the glucose and maltodextrin":** E. S. Chambers, M. W. Bridge, and D. A. Jones, "Carbohydrate Sensing in the Human Mouth: Effects on Exercise Performance and Brain Activity," *Journal of Physiology* 587, pt. 8 (2009): 1779–94.

152 **Through deceptive feedback, the riders were able to increase their perfor-**

mance: M. R. Stone, K. Thomas, M. Wilkinson, A. M. Jones, A. St Clair Gibson, and K. G. Thompson, "Effects of Deception on Exercise Performance: Implications for Determinants of Fatigue in Humans," *Medicine & Science in Sports & Exercise* 44, no. 3 (2012): 534–41.

153 **"a subtle mismatch between the subconscious expectation":** P. C. Castle, N. Maxwell, A. Allchorn, A. R. Mauger, and D. K. White, "Deception of Ambient and Body Core Temperature Improves Self Paced Cycling in Hot, Humid Conditions," *European Journal of Applied Physiology* 112, no. 1 (2012): 377–85.

154 **But the perception of that work remains the same:** J. Swart, R. P. Lamberts, M. I. Lambert, A. St Clair Gibson, E. V. Lambert, J. Skowno, and T. Noakes, "Exercising with Reserve: Evidence That the Central Nervous System Regulates Prolonged Exercise Performance," *British Journal of Sports Medicine* 43, no. 10 (2009): 782–8.

154 **Antidepressants like Wellbutrin and Ritalin have similar effects in heat:** R. Meeusen and B. Roelands, "Central Fatigue and Neurotransmitters, Can Thermoregulation Be Manipulated?" *Scandinavian Journal of Medicine & Science in Sports* 20, suppl. 3 (2010): 19–28.

154 **"The finding that lower RPE scores were recorded":** R. Winchester, L. A. Turner, K. Thomas, L. Ansley, K. G. Thompson, D. Micklewright, and A. St Clair Gibson, "Observer Effects on the Rating of Perceived Exertion and Affect During Exercise in Recreationally Active Males," *Perceptual and Motor Skills* 115, no. 1 (2012): 220.

155 **"Given no effect of mental fatigue on potential motivation":** Samuele M. Marcora, Walter Staiano, and Victoria Manning, "Mental Fatigue Impairs Physical Performance in Humans," *Journal of Applied Physiology* 106, no. 3 (2009): 862.

156 **"Neuroimaging studies are needed to confirm our hypothesis":** Ibid.

157 **same general brain areas identified as possible areas of control:** E. B. Fontes, A. H. Okano, F. De Guio, E. J. Schabort, L. L. Min, F. A. Basset, D. J. Stein, and T. D. Noakes, "Brain Activity and Perceived Exertion During Cycling Exercise: An fMRI Study," *British Journal of Sports Medicine* (June 1, 2013), doi:10.1136/bjsports-2012-091924.

159 **"The winning athlete":** Timothy David Noakes, "Fatigue Is a Brain-Derived Emotion That Regulates the Exercise Behavior to Ensure the Protection of Whole Body Homeostasis," *Frontiers in Physiology* 3 (2012): 8–9, doi:10.3389/fphys.2012.00082.

160 **"It's the brain, not the heart or lungs, that is the critical organ":** Jon Entine, *Taboo: Why Black Athletes Dominate Sports and Why We're Afraid to Talk About It* (New York: Public Affairs, 2000), 14.

160 **"Mind is everything":** Noakes, "Fatigue Is a Brain-Derived Emotion."

Chapter 9: Staying Strong

164 **"Sleep is a reversible behavioral state":** Mary A. Carskadon and William C. Dement, "Normal Human Sleep: An Overview," in *Principles and Practice of Sleep Medicine*, 5th ed., eds. M. H. Kryger, T. Roth, and W. C. Dement (St Louis: Elsevier, 2011).

164 **"all the other indicators one commonly associates with sleeping":** Ibid.

165 **substantial effects on our mood, our mental and cognitive skills:** June J. Pilcher and Allen I. Huffcutt, "Effects of Sleep Deprivation on Performance: A Meta-analysis," *Sleep* 19, no. 4 (1996): 318–26.

165 **certain kinds of athletic tasks are more affected by sleep deprivation:** Shona L. Halson, "Sleep and the Elite Athlete," *Sports Science Exchange* 26, no. 113 (2013): 1–4.

166 **every single player on the team was quicker than before the study had started:** C. D. Mah, K. E. Mah, E. J. Kezirian, and W. C. Dement, "The Effects of Sleep Extension on the Athletic Performance of Collegiate Basketball Players," *Sleep* 34, no. 7 (2011): 943–50.

167 **"If you nap every game day":** Jonathan Abrams, "Napping on Game Day Is Prevalent Among N.B.A. Players," *New York Times*, March 6, 2011, http://www.nytimes.com/2011/03/07/sports/basketball/07naps.html.

168 **Bad news for teams like the Oakland Raiders:** Bill Barnwell, "NFL's Frequent-Flier Phenomenon," Grantland, July 3, 2012, http://www.grantland.com/story/_/id/8125450/packing-miles-hurt-team-overpaying-kicker.

168 **"We theorize that this decline is tied to fatigue":** American Academy of Sleep Medicine, "Fatigue and Sleep Linked to Major League Baseball Performance and Career Longevity," ScienceDaily, May 31, 2013, www.sciencedaily.com/releases/2013/05/130531105506.htm.

169 **The athletes spent more time in bed than the control group:** J. Leeder, M. Glaister, K. Pizzoferro, J. Dawson, and C. Pedlar, "Sleep Duration and Quality in Elite Athletes Measured Using Wristwatch Actigraphy," *Journal of Sports Sciences* 30, no. 6 (2012): 541–5.

169 **A 2011 German study found that 65 percent of athletes:** D. Erlacher, F. Ehrlenspiel, O. A. Adegbesan, and H. G. El-Din, "Sleep Habits in German Athletes Before Important Competitions or Games," *Journal of Sports Sciences* 29, no. 8 (2011): 859–66.

169 **Researchers in South Africa tested a group of runners, cyclists, and triathletes:** L. Kunorozva, K. J. Stephenson, D. E. Rae, and L. C. Roden, "Chronotype and PERIOD3Variable Number Tandem Repeat Polymorphism in Individual Sports Athletes," *Chronobiology International* 29, no. 8 (2012): 1004–10.

170 **advice we can all follow to sleep better:** Shona L. Halson, "Nutrition, Sleep and Recovery," *European Journal of Sport Science* 8, no. 2 (2008): 119–26.

171 **published by two University of North Carolina researchers in 1987:** Michael J. Berry and Robert G. McMurray, "Effects of Graduated Compression Stockings on Blood Lactate Following an Exhaustive Bout of Exercise," *American Journal of Physical Medicine* 66, no. 3 (1987): 121–32.

171 **Part of what makes compression so hard to understand:** Dennis-Peter Born, Billy Sperlich, and Hans-Christer Holmberg, "Bringing Light into the Dark: Effects of Compression Clothing on Performance and Recovery," *International Journal of Sports Physiology and Performance* 8, no. 1 (2013): 4–18.

172 **small boost to performance on short sprints, vertical jump, and time to exhaustion:** Ibid.

172 **Postexercise recovery is where compression shines:** Ibid.

173 **"serve as an ergogenic aid in performance in cold environmental conditions":** Ibid.

173 **A 2012 meta-analysis of studies on athletes and pain:** J. Tesarz, A. K. Schuster, M. Hartmann, A. Gerhardt, and W. Eich, "Pain Perception in Athletes Compared to Normally Active Controls: A Systematic Review with Meta-analysis," *Pain* 153, no. 6 (2012): 1253–62.

174 **"Athletes are frequently exposed to unpleasant sensory experiences":** Ibid., 1259.

174 **A study of players in the 2002 and 2006 soccer World Cup:** P. Tscholl, A. Junge, and J. Dvorak, "The Use of Medication and Nutritional Supplements During FIFA World Cups 2002 and 2006," *British Journal of Sports Medicine* 42, no. 9 (2008): 725–30.

174 **when they've taken ibuprofen before the effort as compared to a placebo:** Jubil L. Young et al., "Ibuprofen and Acetaminophen Have Trivial Effects on Core Temperature and Performance During High Intensity Cycling," presentation at the annual meeting of the American College of Sports Medicine, Indianapolis (May 28–June 1, 2013).

175 **Furthermore, animal studies have shown that taking ibuprofen:** M. Machida and T. Takemasa, "Ibuprofen Administration During Endurance Training Cancels Running-Distance-Dependent Adaptations of Skeletal Muscle in Mice," *Journal of Physiology and Pharmacology* 61 (2010): 559–63.

175 **"such ritual use represents misuse of these potentially dangerous agents":** Stuart J. Warden, "Prophylactic Misuse and Recommended Use of Non-steroidal Anti-inflammatory Drugs by Athletes," *British Journal of Sports Medicine* 43, no. 8 (2009): 549.

175 **The lab, led by Alexis Mauger, has gone on to show:** J. Foster, L. Taylor, B. C. Chrismas, S. L. Watkins, and A. R. Mauger, "The Influence of Acetaminophen on Repeated Sprint Cycling Performance," *European Journal of Applied Physiology* 114, no. 1 (2014): 41–8.

176 **The riders didn't just feel cooler as they exercised:** A. R. Mauger, L. Taylor, C. Harding, B. Wright, J. Foster, and P. C. Castle, "Acute Acetaminophen (Paracetamol) Ingestion Improves Time to Exhaustion During Exercise in the Heat," *Experimental Physiology* 99, no. 1 (2014): 164–71.

177 **then can barely walk two days later:** J. Leeder, C. Gissane, K. van Someren, W. Gregson, and G. Howatson, "Cold Water Immersion and Recovery from Strenuous Exercise: A Meta-analysis," *British Journal of Sports Medicine* 46, no. 4 (2012): 233–40.

177 **The AIS recovery team recently published a review:** Nathan G. Versey, Shona L. Halson, and Brian T. Dawson, "Water Immersion Recovery for Athletes: Effect on Exercise Performance and Practical Recommendations," *Sports Medicine* 43, no. 11 (2013): 1101–30.

177 **contrast water therapy results in less loss of muscular power:** Joanna M. Vaile, Nicholas D. Gill, and Anthony J. Blazevich, "The Effect of Contrast Water Therapy on Symptoms of Delayed Onset Muscle Soreness," *Journal of Strength and Conditioning Research* 21, no. 3 (2007): 697–702.

179 **they also seem to blunt the effectiveness of physical training:** Tina-Tinkara Peternelj and Jeff S. Coombes, "Antioxidant Supplementation During Exercise Training: Beneficial or Detrimental?" *Sports Medicine* 41, no. 12 (2011): 1043–69.

179 **the training doesn't have the same efficiency as it would without the antioxidants:** M. C. Gomez-Cabrera, E. Domenech, M. Romagnoli, A. Arduini, C. Borras, F. V. Pallardo, J. Sastre, and J. Viña, "Oral Administration of Vitamin C Decreases Muscle Mitochondrial Biogenesis and Hampers Training-Induced Adaptations in Endurance Performance," *American Journal of Clinical Nutrition* 87, no. 1 (2008): 142–9.

179 **A scientist at the English Institute of Sport, Jonathan Leeder:** Leeder et al., "Cold Water Immersion and Recovery."

179 **cold-water immersion for the training adaptations of cyclists:** Shona L. Halson, Jason Bartram, Nicholas West, Jessica Stephens, Christos K. Argus, Matthew W. Driller, Charli Sargent, Michele Lastella, Will G. Hopkins, and David T. Martin, "Does Hydrotherapy Help or Hinder Adaptation to Training in Competitive Cyclists?" *Medicine & Science in Sports & Exercise*, published online ahead of print, February 5, 2014, doi: 10.1249/MSS.0000000000000268.

Chapter 10: The Numbers Game

190 **"We assert that 'dominant' interior defense can manifest":** Kirk Goldsberry and Eric Weiss, "The Dwight Effect: A New Ensemble of Interior Defense Analytics for the NBA," paper presented at the MIT Sloan Sports Analytics Conference (March 1, 2013), www.sloansportsconference.com/wp-content/uploads/2013/

The%20Dwight%20Effect%20A%20New%20Ensemble%20of%20
Interior%20Defense%20Analytics%20for%20the%20NBA.pdf.

190 **allowing shooters to make 63 and 53 percent of their shots:** Ibid.

191 **"This is what we call the 'Dwight Effect'":** Ibid., 5.

Chapter 11: Athlete's Little Helper

203 **95 percent of high school students admit to some form of cheating:** Jeremy P. Meyer, "Students' Cheating Takes a High-tech Turn," *Denver Post*, May 27, 2010, http://www.denverpost.com/news/ci_15170333.

203 **engaged in "serious cheating" such as cheating on a test or plagiarism:** Donald L. McCabe, Linda Klebe Treviño, and Kenneth D. Butterfield, "Cheating in Academic Institutions: A Decade of Research," *Ethics & Behavior* 11, no. 3 (2001): 219–32.

205 **only two of the 250 respondents, just 0.8 percent, said they'd do so:** J. M. Connor and J. Mazanov, "Would You Dope? A General Population Test of the Goldman Dilemma," *British Journal of Sports Medicine* 43, no. 11 (2009): 871–2.

205 **about the same low rate found in the general population study:** James Connor, Jules Woolf, and Jason Mazanov, "Would They Dope? Revisiting the Goldman Dilemma," *British Journal of Sports Medicine* 47, no. 11 (2013): 697–700.

206 **understand the prevalence of doping and the attitude of athletes toward it:** L. Whitaker, J. Long, A. Petróczi, and S. H. Backhouse, "Using the Prototype Willingness Model to Predict Doping in Sport," *Scandinavian Journal of Medicine & Science in Sports* (2013), doi:10.1111/sms.12148.

207 **the athlete was simply held out of the international competition:** Werner W. Franke and Brigitte Berendonk, "Hormonal Doping and Androgenization of Athletes: A Secret Program of the German Democratic Republic Government," *Clinical Chemistry* 43, no. 7 (1997): 1262–79.

207 **how far a country would go for sporting success:** Ibid.

208 **"At present anabolic steroids are applied in all Olympic sporting events":** Ibid., 1264.

208 **"In numerous women the prevailing administration of anabolic hormones":** Ibid., 1274.

209 **countries that we know had systematic doping programs:** Charles E. Yesalis and Michael S. Bahrke, "History of Doping in Sport," *International Sports Studies* 24, no. 1 (2002): 42–76.

210 **still were 3 percent faster than their first time trial:** J. Durussel, E. Daskalaki, M. Anderson, T. Chatterji, D. H. Wondimu, N. Padmanabhan, R. K. Patel, J. D. McClure, and Y. P. Pitsiladis, "Haemoglobin Mass and Running Time Trial Performance After Recombinant Human Erythropoietin Administration in Trained Men," *PloS One* 8, no. 2 (2013): e56151, doi:10.1371/journal.pone.0056151.

212 **a ridge that runs through the two countries at high altitude:** Wilber and Pitsiladis, "Kenyan and Ethiopian Distance Runners."

212 **higher altitudes didn't lead to increased adaptation:** Robert F. Chapman, Trine
Karlsen, Geir K. Resaland, R. L. Ge, Matthew P. Harber, Sarah Witkowski, James
Stray-Gundersen, Benjamin D. Levine, "Defining the 'Dose' of Altitude Training:
How High to Live for Optimal Sea Level Performance," *Journal of Applied Physi-
ology* 116, no. 6 (2014): 595–603.

216 **"For example: Hamilton's natural hematocrit is typically 42":** Tyler Hamilton
and Daniel Coyle, *The Secret Race: Inside the Hidden World of the Tour de France*
(New York: Bantam, 2012), 62.

219 **When the researchers treated female mice with testosterone:** I. M. Egner, J. C.
Bruusgaard, E. Eftestøl, and K. Gundersen, "A Cellular Memory Mechanism
Aids Overload Hypertrophy in Muscle Long After an Episodic Exposure to Ana-
bolic Steroids," *Journal of Physiology* 591, pt. 24 (2013): 6221–30.

220 **"Even though the direct effect of doping substances":** Ehlert, Simon, and
Moser, "Epigenetics in Sports," 102.

Chapter 12: The Limits of Performance

222 **move efficiently and run long distances without reaching exhaustion:** Daniel
E. Lieberman and Dennis M. Bramble, "The Evolution of Marathon Running:
Capabilities in Humans," *Sports Medicine* 37, no. 4–5 (2007): 288–90.

223 **"In all measurable Olympic contests from five different disciplines":** Geoffroy
Berthelot, Valérie Thibault, Muriel Tafflet, Sylvie Escolano, Nour El Helou, Xavier
Jouven, Olivier Hermine, and Jean-François Toussaint, "The Citius End: World
Records Progression Announces the Completion of a Brief Ultra-Physiological
Quest," *PloS One* 3, no. 2 (2008): e1552, doi:10.1371/journal.pone.0001552.

224 **has come simply from the growth in the world's population:** Scott M. Berry,
"A Statistician Reads the Sports Pages: One Modern Man or 15 Tarzans?" *Chance*
15, no. 2 (2002): 49–53.

226 **Using some fancy physics modeling:** Ajun Tan and John Zumerchik, "Kinemat-
ics of the Long Jump," *Physics Teacher* 38, no. 3 (2000): 147–9.

230 **"The catcher held the ball for a few seconds":** Pat Jordan, "The Wildest Fastball
Ever," *Sports Illustrated*, October 12, 1970, http://sportsillustrated.cnn.com/vault/
article/magazine/MAG1084175/1/index.htm.

231 **"selection for throwing as a means to hunt":** N. T. Roach, M. Venkadesan, M. J.
Rainbow, and D. E. Lieberman, "Elastic Energy Storage in the Shoulder and the
Evolution of High-Speed Throwing in Homo," *Nature* 498, no. 7455 (2013):
483–6.

233 **Baseball injury expert Will Carroll reported in 2013:** Will Carroll, "The

Alarming Increase in MLB Pitchers Who've Had Tommy John Surgery," Bleacher Report, July 17, 2013, http://bleacherreport.com/articles/1699659-the-alarming -increase-in-mlb-pitchers-whove-had-tommy-john-surgery.

233 **"There may be an outlier, one exception here or there":** Barry Bearak, "Harvey's Injury Shows Pitchers Have a Speed Limit," *New York Times*, September 16, 2013, http://www.nytimes.com/2013/09/17/sports/baseball/harveys-injury-shows -pitchers-have-a-speed-limit.html.

233 **"Paleolithic hunters almost certainly threw less frequently than modern athletes":** Roach et al., "Elastic Energy Storage in the Shoulder," 486.

234 **"I didn't see the ball until it was behind me":** Steve Henson, "Chapman Throws Fastest Pitch Ever Recorded," Yahoo Sports, September 25, 2010, http://sports .yahoo.com/news/chapman-throws-fastest-pitch-ever-074900166—mlb.html.

235 **the Preakness and Belmont reached plateaus in 1971 and 1973, respectively:** Mark W. Denny, "Limits to Running Speed in Dogs, Horses and Humans," *Journal of Experimental Biology* 211, pt. 24 (2008): 3836–49.

235 **speeds reached a plateau between 1966 and 1971:** Ibid.

235 **"Despite intensive programs to breed faster thoroughbreds":** Ibid., 3842.

236 **just 0.33 percent slower than Denny's predicted absolute record of 2:14:58:** Ibid.

INDEX

weight training, 48
Weyand, Peter, 9
whole vs. part training, 100–101
Wiggins, Bradley, 5, 70, 71
Wilber, Randy, 167, 210–15
Wilkins, Dominique, 185–86
Williams, Ted, 230
Woods, Tiger, 96, 97, 124, 135
World Anti-Doping Agency (WADA),
    201–3, 204, 205, 210, 215, 218
wrestling, 84, 119–120

Youkilis, Kevin, 198
young athletes
    backgrounds of, 69–72
    developmental curve in, 73, 84, 85–86
    injuries among, 86
    late bloomers, 64–66
    optimal events for, 74–75
    and relative age effect, 66–68
    and size of community, 68–69
    specialization among, 86, 87
    and talent identification, 75